World Soils Book Series

Aims and Scope

The *World Soils Book Series* brings together soil information and soil knowledge of a particular country in a concise and reader-friendly way. The books include sections on soil research history, geomorphology, major soil types, soil maps, soil properties, soil classification, soil fertility, land use and vegetation, soil management, and soils and humans.

International Union of Soil Sciences

More information about this series at http://www.springer.com/series/8915

Nicholas Yassoglou · Christos Tsadilas
Costas Kosmas

The Soils of Greece

Springer

Nicholas Yassoglou
Agricultural University of Athens
Athens
Greece

Costas Kosmas
Agricultural University of Athens
Athens
Greece

Christos Tsadilas
Institute of Industrial and Forage Crops
Hellenic Agricultural Organization
Larissa
Greece

ISSN 2211-1255 ISSN 2211-1263 (electronic)
World Soils Book Series
ISBN 978-3-319-53332-2 ISBN 978-3-319-53334-6 (eBook)
DOI 10.1007/978-3-319-53334-6

Library of Congress Control Number: 2017930611

Printed on acid-free paper

This Springer imprint is published by Springer Nature
The registered company is Springer International Publishing AG
The registered company address is: Gewerbestrasse 11, 6330 Cham, Switzerland

“Φασί γάρ τόν μέλλοντα ὀρθῶς γεωργήσειν τῆν φύσιν χρῆναι πρῶτον τῆς γῆς εἰδέναι”

“It is said that he who intends to correctly exercise agriculture, he needs first to know the nature of the land”.

Xenophon’s apofthegm, Oikonomikos, XVI

This book is dedicated to the numerous eponymous and anonymous persons contributed in any way in the mapping, classification and research on the soils of Greece.

Preface

From ancient times up to date, agriculture has been the principle use of soil. It contributed in 2008 to the Greek economy at a percentage of 6.5% of the gross domestic product (GDP). About 3.8 million ha from the total area of the country (13.2 million ha) is agricultural land not including pasture (Ministry of Agriculture 2008). In addition, agriculture significantly contributes to other economic sectors such as industry, construction, wholesale and retail trade, and other services, although recent studies showed that agriculture followed independent course in relation to the other economic sectors not affecting them seriously (Kyrkilis et al. 2014). However, with the implementation of the new Common Agricultural Policy in European Union after 2006, agriculture started to decline and the agriculture outputs are at the present lower than in 1970, contributing to the GDP less than 2% (Kyrkilis and Semasis 2015). From the previously mentioned data, it becomes clear that soil, as one of the most important factors of the agricultural production, has played and continues to play an important role in the evolution of economy and society in Greece. Soils of Greece are cultivated for thousands of years intensively and under unfavorable conditions, as the Mediterranean environment has been seriously degraded, loosing much of their organic matter and being strongly eroded in the sloping areas. As it was pre-mentioned, agriculture constitutes a basic sector of the economy of Greece. Therefore, the maintenance of soil quality and soil sustainability, on which agricultural production is mainly based, constitutes a crucial factor for the economic and social development.

This book is consisted of 11 chapters as follows:

In Chap. 1 (Introduction), an extensive reference is made on the history of Greek agriculture in ancient times and the contribution and significance of soils in its performance. In addition, genetic trends and development of soils are described as they proceeded under the combined influence of the basic soil-forming factors, and the main threats of the soils of Greece are presented.

In Chap. 2, the characteristics of the main soil-forming factors such as climate, parent material, relief, the biotic factor, and the human actions are described.

In Chap. 3, the soil mapping and classification of the soils of Greece as they were produced by the respective authorities of the Ministry of Rural Development and Foods, the research institutes, and the universities based on the soil mapping and classification systems in use in the country and, internationally, are presented. In addition, a delineation of the soils according to their principal characteristics such as slope and depth through maps and tables is presented.

Chapter 4 describes the main characteristics and properties of the soils in mountainous and hilly areas formed under the strong influence of the geomorphology and surface geology and the climate. It is noticed that these soils cover more than 60% of the total country area being a significant soil resource for the Greek agriculture.

In Chap. 6, the soils developed on argillic deposits and conglomerates, which are extended in the whole country covering areas where geological and eroded soil materials were transferred from the original places and have been deposited on lower lands, are described.

Chapter 7 presents the inorganic soils that have been formed on recent lake beds and river deltas and usually coexist with recent alluvial deposits as well as organic soils.

In Chap. 8, characteristics and properties of soils developed on alluvial deposits which are the most productive soils of the country are presented in detail. These soils are distinguished in those formed on Holocene deposits and Pliocene deposits.

In Chap. 9, the management groups of the soils developed on alluvial deposits are described in detail, since these soils are the most important agricultural soils in Greece.

Chapter 10 is referred to the land degradation and desertification already happened or is about to occur due to the Greek conditions favoring these phenomena taking place from the ancient times to date. Measures of combating desertification are also discussed.

In the last Chap. 11, a brief presentation of the research carried out so far and its results is made. It includes the basic fields of soil science such as soil mapping and classification, soil genesis, soil degradation and protection, soil fertility and fertilization, soil physicochemical characteristics, soil quality, soil and water pollution, and waste management in relation to soils.

Athens, Greece — Nicholas Yassoglou
Larissa, Greece — Christos Tsadilas
Athens, Greece — Costas Kosmas

Acknowledgements

Firstly, the authors would like to thank Dr. Alfred. E. Hartemink, who proposed the publication of this volume. The data, which this book are based on, come from several sources of the both public and private sectors including the Ministry of Agriculture, the National Agricultural Research Foundation of Greece (NAGREF), Agricultural University of Athens, Aristotle University of Thessaloniki, the former private Research Institute "Kenellopoulos," and the publications of many scientists who dedicated most of their professional life to the study of the soils of Greece. To all of them either referred or not in the book, we express our warm thanks. Thanks are also expressed to the board of directors of the Hellenic Agricultural Organization DEMETER (former NAGREF) for the permission to use soil data from soil survey studies carried out by its institutes.

Contents

About the Authors

Nicholas Yassoglou is a forester with a M.Sc. from University of Thessaloniki and a Ph.D. from Michigan State University. He is an Emeritus Professor of soil science in the Agricultural University of Athens. He is considered as a pioneer in the fields of modern Greek soil survey and land evaluation. Establishment and organization of soil units and the development of Higher Education. He was the chairman of the Greek Committee for Combating Desertification for a decade, and under his coordination, the National Action Plan was compiled. He serves as a representative of the Greek team of the European Topic Centre on Soil for a long period. He was a member of the Advisory and Scientific Committees of the European Bureau and coordinated its ad hoc working group on soil erosion.

Christos Tsadilas is an agronomist with an M.Sc. and Ph.D. from the Agricultural University of Athens. He has been the director of the Institute of Soil Survey and Mapping of the National Agricultural Research Foundation of Greece (NAGREF) and at the present director of the Institute of Industrial and Forage Crops of the Hellanic Agricultural Organization DEMETER. He has worked for more than 30 years in the fields of soil survey, soil chemistry and fertility, soil pollution and restoration, and waste management. He was a founding member of the Hellenic Soil Science Society and its president for more than five years. He has participated in several research projects both national and European and published more than 250 papers in international journals and conference proceedings, and chapters in books.

Costas Kosmas is an agronomist with an M.Sc. from Agricultural University of Athens and a Ph.D. from Purdue University (Indiana, USA). He is an Emeritus Professor of soils in the Agricultural University of Athens. He has over 30 years experience in management of land resources and the environment mainly in agricultural ecosystems. He has participated as research leader or project coordinator in more than 24 educational and research EU and Greek projects, related to soil survey, soil management, water pollution, soil erosion, land degradation, land desertification, and environmental protection. He was served as a member and chairman of the Greek National Committee for Combating Desertification.

List of Figures

List of Tables

1 Introduction

1.1 Land use in Ancient Greece

Agriculture has been a significant sector of the Greek economy from the ancient times to date. Agriculture in ancient Greece was praised and clearly contrasted with other economic activities. Already since 390 B.C., the Athenian historian Xenophon in its very famous work Economics (in Greek "Oikonomikos") suggested that agriculture contributes substantially to the growth of the whole economic system, providing to the people and the societies not only economic wealth but also stimulation of the spiritual well-being. The same historian expressed the view that agriculture constitutes the foundation of the economy, claiming that when all those related to agriculture go well, then all the other arts do well, but when for any reason earth is not cultivated, then necessarily all the arts decline. In another work entitled *Sources* (in Greek "Poroi"), Xenophon expressed for the first time the very famous *law of diminishing returns*. In the earliest literary sources, Homer and Hesiod described agriculture as one of the foundations of civilized life, closely connected with family life. Land was the basis of the material wealth of an *oikos* (home) which was the central institution in the Homeric world, before the subsequent emergence of the *polis* (city). Greece was an agrarian society, as every society was before the Industrial Revolution. Therefore, land was the most secure resource because landownership provided direct control of the means of production. This of course does not mean that all cities such as Athens were exclusively agrarian. In fact, there were two groups of the population, i.e., one that constituted the great majority which was involved in agricultural activities and, the other, a small proportion of the population often more creative and energetic, capable to find enough capital, dealt with other economic activities such as industry or shipping. However, most general surveys of Greek history are urban in origin. The Greeks appear as congregations of bards, poets, and philosophers, an approach that is misdirected as Hanson (1995) stated. He claimed that the "Western values," which is widely acceptable that originate from ancient Greeks, own to "Other Greeks," i.e., the outside of the *polis* in countryside who were dealing exclusively with agriculture. Information on agriculture's organization in ancient Greece (Attica, Peloponnese, islands, Thessaly, Macedonia, Thrace), on its significance, and on the role of soil and land, comes mostly from Hesiod (*Works and Days*), Xenophon (*Economics*), and Theophrastus (*Historia Plantarum*).

The climate of the region of ancient Greece was Mediterranean, which is characterized by mild and rainy winters and dry hot summers, unevenly distributed rainfall coming frequently as torrents during winter months, while summers are very/quite/rather dry. Considerable climatic variation occurs between East and West Greece with west part of Greece having the highest amount of rainfall with the highest amount of precipitation in Pindos range. Except for the high mountains, only few areas suffered from frosts. In an oversimplification, the Mediterranean zone is defined as the area of the olive's cold tolerance (Grove and Grackham 2003). Both Mediterranean environment and ecosystems are fragile and resilient. Since the area is inhabited and exploited for many thousands of years, it is very difficult to find natural environments unaffected by human activities. The main perennial natural vegetation consists of oak, chestnut, and pine forests which are the "climax vegetation as well as maquis and garrigue; the short-lived perennials such as mints, thyme, oregano, caper, bryony and smilax; and the common annuals vetches, wild carrot, wild fennel, grasses, and bulbs (crocus, colchium, asphodel, squill, cyclamen). These vegetational communities are unstable and easily damaged; however, they possess excellent capacity of recovery. Concerning the argument on the climatic change, it would be said that it could not be postulated that no climatic changes occurred in the time of antiquity up to date. In contrast, investigations at Philippi in Macedonia and Lake Kopais in Boetia testified that the climate was stable from Palaeolithic age (before 10,000 BC) to the recent time and considerable deforestation of the virgin oak forest was realized/occurred during Bronze age (3300–1200 BC)

N. Yassoglou et al., *The Soils of Greece*, World Soils Book Series,
DOI 10.1007/978-3-319-53334-6_1

(Isager and Skydsgaard 1995). Climatic changes may happened within the historical periods, but there is no evidences that the climate in ancient Greece was essentially different than today. There are historical evidences that the environment in the region of ancient Greece has been destructed from human and natural causes and recovered. However, we have to keep in mind that technology used in ancient Greek culture was not so advanced to be able to damage the environment like in the present era (Foxhall 2006).

Land was used in every form, such as for agriculture, and cultivation of vines, olive trees, and vegetables. The average land of a household in ancient Greece was 3–5 ha as it is in modern Greece (Amemiya 2007). Landownership and citizenship were closely linked. In most cities, only citizens could own land and landowners could be citizens. As it appears in several Greek texts of the classical period (fifth to fourth centuries BC) such as Plato's *Critias*, landownership and farming were assigning a high moral and social value to the citizens. From this source, it comes up that farming was central to the lives of most of the Greeks, and financial profit and moral gain were interlaced. Furthermore, techniques and principles related to farming were quite familiar to all people (Foxhall 2006). Cultivation of soils in sloping areas was more difficult than in flat areas and had to be done in a way that would protect the soil from the harmful effect of rainwater. Two practices are referred in the literature concerning this issue. The first one was to construct terraces in order to create a flat area suitable for cultivation, to decrease the speed of runoff from the rains, and thus to hold soil in place, protecting it this way from erosion. The second way was to dig trenches around the trees such as basins, in which water from rainfall would be kept benefiting the trees. In cases where drainage was a problem, the trenches were connected with ditches across the slope. As Theophrastos argues in the fourth century, the repeating digging around the trees had also other beneficial results, such as removal of weeds which would compete for nutrients and water, keeping the soil dry and crumbly, reducing thus loss of water through evaporation because of the interruption of capillary action. In sloping areas, this time-consuming work was performed by the slaves. In flat areas, cultivation consisted of plowing in summer and winter. The creation of a proper seedbed was achieved through the removal of woody weeds and the beneficial snow effect on soil during winter.

The main crops grown in ancient Greece were grains, a term used for wheat (*sitos*), barley (*krithe*), millet (*kegchros*), legumes (beans, lentils, chickpeas, vetch, lathyrus), grapes, olives, and other fruit trees such as pears, pomegranates, apples, quinces, medlars, figs, almonds, and walnuts, as well as garden vegetables such as marrow, cucumbers, leeks, onions, garlic, and chicory. Surprisingly, vegetables cultivated nowadays in the Mediterranean, such as tomato, potato, and peppers, were unknown in classical times. The productivity of the land was very low compared to yields achieved these days. For example, the productivity for wheat in ancient Greece is estimated to range at 465–620 kg/ha and for barley at 765 kg/ha (Amemiya 2007), or according to Theophrastos at 750–1000 kg/ha per biennium in alternate year fallow. Probable maximum yield for oil was 3.4 kg/tree per biennium (Foxhall 2007). This low productivity of agriculture reflects the level of the factors affecting it, including the yield capacity of the genetic resources, the equipments used, the existing supportive structures, and the farming systems. However, for some tree crops such as fig trees, the productivity was very high amounting to 50–100 kg dried figs per small and less productive and large fig trees, respectively. This is noticed because dried figs were a major staple especially of poor people or slaves.

The farming tools used in ancient Greece, as they are described by Theophrastos and Hesiod, were simple instructions made of wood and iron. Hesiod in his poem "Works and Days" described the plow used for soil cultivation, a tool that is used identically until recently, called *Hesiod's plough*. This was basically used for plowing shallow soils to sow arable crops such as cereals. Other farming tolls were mattocks for digging, sickles for cereals harvesting, winnowing forks for separating straw from grains in cereal threshing, pruning knives, adzes, and threshing sledges.

Based on the epigraphies and archaeological documents, we know that a great variety of installations were found. Stone-paved threshing floors have been found in Euboia and in Attica. On steep lands, terrace wall constructions have been used, especially for olive cultivation or even for cereals and many other crops for expansion of farmers into sloping marginal lands, in periods of economic buoyancy. Another important kind of installations in agriculture was the one related to drainage and irrigation such as wells, cisterns, and water lifting devices, which significantly improved the cultivation potential and the value of land by facilitating the planting of crop trees and small-scale irrigation (Foxhall 2007). One of the most significant functions of these constructions was soil protection from erosion.

It was quite clear from the ancient times that availability of water was the primary limiting variable/factor for tree growth. According to Theophrastos (372–287 BC), in the absence of adequate moisture, overconcentration of soil nutrients, due to the addition of high amount of manure, may harm the trees, especially the young ones. Evidently, soil management in ancient Greece was basically aiming to save and utilize the available moisture. Irrigation was very important for agriculture due to the shortage of water in the Mediterranean environment of Greece. The importance of water to Greeks is reflected to the fact that in *Plato's Laws,* irrigation was in common use. The shortage of water necessitated the development of various water management systems, including the collection in storage cisterns. Such systems were initially developed at the Minoan times (3200–1100 BC) and continued up to the Hellenistic period (323–30 BC). In Minoan Crete, a technology related to collecting, transporting, and using water was highly developed. Rainwater was collected, transferred, and stored in cisterns from several roofs and yards of buildings and used for irrigation. Hydraulic structures associated with the collection of rainwater were found in several places in Crete, such as Knossos, Phaistos, and Tylisos (Antoniou et al. 2014). A similar practice is also met today in rural areas of Greek islands.

Agricultural activities are closely related to animal husbandry, since from one side animals are exploited resources coming from plant growth on fallow or uncultivated lands and from the other side agricultural by-products such as manures produced by animals are used for plant growth. The animals bred were mostly sheep and goats, while cattle were very difficult to be maintained under the dry Mediterranean conditions. Chicken were also bred everywhere in Greece in the ancient times. The regular practice was the movement of the flocks from lowland grazing in winter to upland grazing in summer, which to some degree is also met nowadays.

From the above-mentioned information, it merges that agriculture played a fundamental role in the ancient Greek civilization, being the basic sector of its economy. Thus, soils, through their function as medium of plant growth, offered an irreplaceable service to the development of ancient Greek culture. Cultivated for thousands of years, the soils received an enormous pressure, and as a result, they were substantially degraded. There is no clear historical evidence from which the degree of soil degradation can be estimated. Surprisingly, although the ancient Greek historians, philosophers, poets, and writers expressed their opinion on so many topics, they did not analyze adequately the methods of land management from which Greek economy was seriously affected. So, regarding theuse of land, we have to employ modern soil science methods and approaches to reach sound conclusions. Farming tools used and methods of soil cultivation were so sparse and crude that it is reasonable to accept that land damage was quite slower and less than that of the modern times, where mechanization and general intensification of land management were applied. The factor that promoted land degradation was the sloping landscape of most of the agrarian areas, which favored/promoted/led to soil erosion. The basic land management measure to mediate it was the construction of terraces or trenches around the trunk trees in order to decrease the surface water flow rate and favor water movement downward in soil profile. Another serious reason that might have affected land quality was the frequent wars in ancient Greece. During the wars, a number of actions had to take place that were inevitably destroying agriculture and affecting land quality. For example, destruction of grain and cutting down of olive trees and vines were considered necessary measures, during wartime. This fact had serious broader consequences, creating economic crises which in turn affected land management, due to permanent agricultural devastation. Since there are no safe historical data, the degree of the impacts on the land has been estimated in different ways. Some authors believe that warfare in ancient Greece had a low impact on agriculture and consequently on land (Hanson 1998), while other wars had serious consequences on the land and on the overall economy of the country (Dale and Carter 1955).

1.2 Genetical Trends

Soils of Greece, as all soils are a part of the local ecosystem. It is affected by and contributes to forces operating in it. Its physical, chemical, and biological properties and ongoing processes depend on Jenny's (1941) soil-forming factors: parent material, climate, topography, organisms, and time. According to the same author (Jenny 1980), these factors are expressed by the extended equation:

$$l, v, a, s = f(cl,\ o,\ r,\ p,\ t, \ldots)$$

where l any total system properties, v any vegetation properties, a any animal properties, cl climate, o fauna and flora, r topography and water table, p parent material, and t age or time.

The soil-forming factors, appearing in the right side of Jenny's equation, were considered as the five independent parameters controlling soil properties. However, the distinction is not clear for some of them. Fauna and flora, for example, depend almost on all other four and on the dependent soil properties. Climate, on local scales, depends on aspect and in some cases on the parent rock. Man's actions on the ecosystem cause, in relative short time, dramatic changes on the soil system, disrupting the natural course of soil genesis processes. In many cases, they promote retrogressive genetic processes, leading to the retreat of soil profiles from the relative stable equilibrium state of theclimax soil profile developed in natural undisturbed ecosystem.

Soils of Greece, as many Mediterranean soils, have suffered intensive human interference for long periods of time. The most prominent effect has been the loss of successive soil layers or horizons from the soil profiles onsloping land caused by the removal of the protective vegetative cover and the subsequent erosion of the soil surface.

Soil landscapes on mountainous and hilly lands with slope gradients greater than 10% form retrogressive sequences, with members at various degrees of deviation from theclimax soil of a particular ecosystem. The morphology and the physical, chemical, and biotic properties of theclimax soil and of its retrogressive associates reflect the parent material, the physiography, the climate, and intensity and duration of the anthropic interference. For example, in the case of a climaxLuvisol (Alfisol), soil profile has lost the characteristic argillic horizon and has retained the underlying altered layer qualified as cambic horizon, and itsretrogressive erosional associate could be Cambisol (Inceptisol). If it has lost both horizons, it would be a Regosol (Entisol) whereas, if the remaining soil material overlies bedrock and its depth is less than 10 cm, it would be a Leptosol (Lithic Orthents).

The course and the intensity of the genesis process, as well as the properties of theclimax soils of each association in the above landscapes, depend on five soil factors, but most heavily on the parent material and the parent rock, the relief, and the climatic zone. The latter is expressed effectively by the vegetation zone. The genetic status and the properties of the retrogressive erosion associates of each association undergo rather fast changes and depend primarily on the parent material, the relief, and the intensity of the human interference. The latter refers mainly to the removal of the protective vegetation by excessive lumbering, clear cutting for unsustainable farming and pastures, and overgrazing.

Soils on gently sloping TertiaryandQuaternary terraces have not suffered extensive erosion and retain their diagnostic horizons. However, many of them, on Tertiary and Quaternary marly deposits, have lost their surface Mollic, Epipedon through either erosion or long-time cultivation. In addition, recent alluvial planes have been subjected to sedimentation by flooding with soil materials eroded from the upland hilly areas. The dominant soil-forming factors affecting the genetic status of these soils are as follows: the composition of the parent material, the age, as estimated from their geomorphic position in the terraced catenas and from the intensity of the human interference. The climate in these landscapes has been constant throughout the country for some thousands of years (Wright 1972) and has not produced significant differentiation among the soils.

Landscapes, at elevation below 30 m from the sea level or from the inland flood plains, are dominated by recent alluvial deposits. The soils do not exhibit genetic horizons, other than weakly formed cambic and calcic horizons in imperfectly drained sites. Their properties are controlled by (a) the texture and composition of their successive alluvial layers constituting their profiles, (b) the anthropic interference through cultivation drainage and irrigation, (c) naturally or human-induced processes of salinization, and (d) the geomorphic position in the alluvial plain, which influences both the texture of the deposits, the hydromorpic, and the salic characteristics of the soils. Climate, particularly rainfall, induces periodical hydromorphic conditions to the fine textured soils in the subhumid zone.

Soil taxonomic units have been grouped into soil associations, the members of which are:

(i) Retrogresive erosional sequences of the sloping mountain and hilly lands with dominant soil-forming factors such as the parent material, the climatic zone, and the anthropic interference.
(ii) Geomorphic time catenas on the terraced Tertiary and Quaternary deposits, with dominant soil-forming factors age and composition of parent material.
(iii) Textural, hydromorphic, and salic sequences in recent alluvial deposits, with dominant soil-forming factor the texture and composition of the parent material.

Thetaxonomic units have been classified according to the FAO (1998) and Soil Survey Staff (1992). They also were used as soil mapping units of the 1:1,000,000 and the 1:500,000 scale soil maps of Greece.

1.3 Main Threats of the Soils of Greece

Agricultural land worldwide is seriously threaded by degradation due to climatic change and mismanagement. In extreme cases, land degradation leads to desertification, i.e., a non-reversible step of that. Desertification is considered as the greatest threat of land resources which led to the United Nations to sign the United Nations Convention to Combat Desertification (U.N.C.C.D.) in 1974. Since the Mediterranean environment is favorable to land degradation, in the convention a special appendix was devoted to the Mediterranean countries (Annex IV). In Greece, the convention legislated and entered into force in 1997 with the law 2468/97 which imposed the implementation of a National Program of Action through which an appointed committee prepared a plan of measures to combat desertification. At the same time, the European Union funded a series of research projects to obtain a better knowledge on the desertification progress and processes through which a substantial advancement was made on these issues.

A number of indicators were developed through the recent research in Europe to present the environmental sensitivity to desertification, grouped in four categories including soil, climate, vegetation, and management (Kosmas et al. 1999). Among them, soil and management factors have the greatest effect ondesertification.

Soils of Greece were developed on landscape consisting mostly of alternating mountains and hills and small valleys, while parent rocks very commonly are revealed. It is estimated that about 50% of the country has a slope >10% and only 36% a slope less than 5%. The geological background consists of limestone, marls, and volcanic rocks. Soils of Greece on limestone are very often shallow, with high water permeability and thus no quite favorable for supporting vegetation. This way, great areas lied on Mesozoic limestone were desertified. In contrast, soils formed on marls usually have an adequate depth to grow agricultural crops, but they also are characterized as sensitive to desertification, due to the dry environment because of the low rainfall. On acid igneous and volcanic rocks, coarse textured soils were formed which are medium sensitive todesertification (G.N.C. C.D. 2000). Based on the sensitivity indices proposed by Kosmas et al. (1999), the soilquality index of the soils of Greece is as follows: 6% high, about 64% moderate, and 27% low (Karamesouti 2011).

The respective values of the index of soil protection from erosion show that more than 50% have from very low to moderate soil protection.

Although in areas with natural characteristics favoring land degradation, such as intense relief and Mediterranean

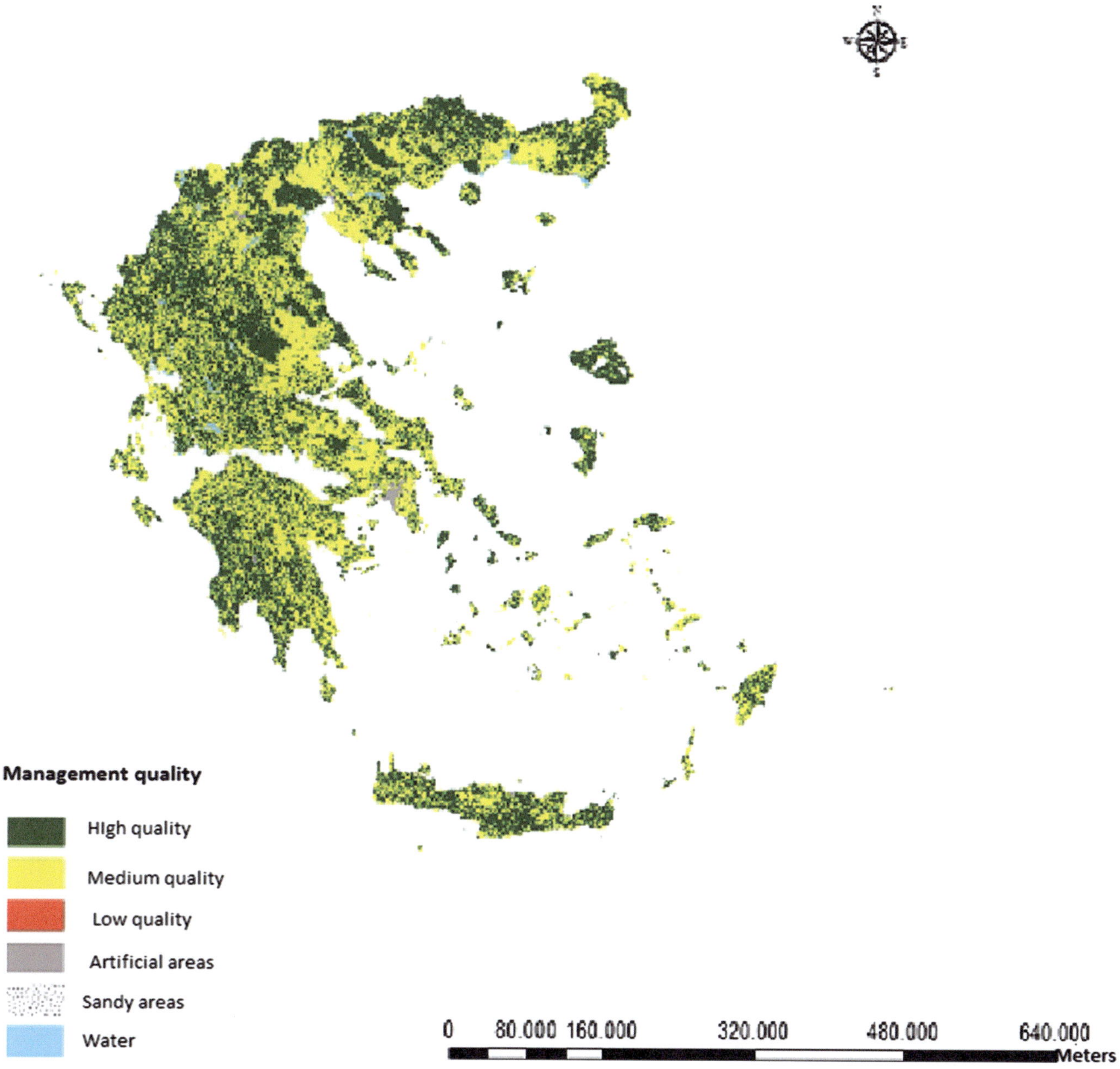

Fig. 1.1 Quality management map of land of Greece (Karamesouti 2011 and Kosmas personal data)

climate, a very careful management is needed to avoid land degradation; in general, land was not well managed and thus favored land degradation. Practices that led to these results were forest clearance, deforestation, forest fires, overgrazing ofsloping lands, and change of pastures to agricultural land. This land mismanagement resulted in an increase of surface water flow, a continuous increase of water consumption and on overexploitation of ground water in flat areas. The quality management index of the land of Greece according to pre-mentioned system (Kosmas et al. 1999) suggests that about half of the percentage is moderate and the rest half is high (Fig. 1.1).

2 Soil-Forming Factors

Soil-forming factors affect not only the course of the genesis processes but also the present and future performance of the soil systems. A brief evaluation of these factors and of their influence on the soils of Greece is presented in this section. The properties described below refer to climax undisturbed soil profiles developed under natural vegetation, which has not been disturbed to a degree that would lead to the truncation of the soil profile by erosion. It must be pointed out, however, that these conditions do not prevail in the sloping landscapes of Greece.

2.1 Climate

Greece is located in the northeastern section of the Mediterranean between the 34th and the 42nd parallels and between the 19th and 29th longitudes of the Northern Hemisphere. Therefore, in general terms, the country is dominated by the Mediterranean type of climate with mild and wet winters and warm and dry summers. However, the extremely irregular topographic configuration, characterized by from north to south significant elevation differences, which causes considerable variability of climatic types within relatively short distance. Mountain ranges dissecting the mainland from north to south and the west–east range in northern Greece affect wet climatic zones in the west and the north sections of the country, whereas the eastern coastal and low-elevation hilly provinces are generally drier. Areas at high elevations along the northern borders of the country have Central European climate.

The country has two climatic seasons: (a) the cold and rainy from the middle October to the end of February with average minimum temperatures 5–10 °C in the coastal and 0–5 °C in inland areas. In northern areas, minimum temperatures are below zero. The warmest period is from July 20 to August 10 with average maximum temperatures ranging between 29 °C and 35 °C.

Greece has been subdivided by Mariolopoulos (1938) into five climatic regions:

(a) The mountainous region
(b) The north or continental region
(c) The Ionian or marine Mediterranean region
(d) The Aegean or inland Mediterranean region
(e) The Cretean or desert-like Mediterranean region

(a) The Mountainous Region

The mountainous region consists of Mount Olympus and the long mountainous range, which runs from north to northwest and divides the mainland into two parts. It includes all the high mountains, and its lands have elevation higher than 1000 m in northern sections and 1200 m in southern. It occupies 16,446 km^2. The climatic parameters of this region are as follows:

Mean temperature is 8–10 °C, which decreases with elevation by 0.6–0.8 °C/100 m. Mean annual rainfall ranges between 1400 and 2220 mm, number of rainy days greater than 100 frost days per year more than 100.

The soil moisture regime in this region is udic. The leaching of soil bases is extensive resulting in dystric and occasionally on acid rocks ultic characteristics. The dominant diagnostic horizon of the climax soil is the cambic. The high acidity of the soil materials prevents illuviation of clay and the formation of the argillic horizon, except on limestone, where climax soils frequently develop argillic horizons.

(b) The Continental Region of northern Greece

The area of this region is 46,900 km^2 and includes large sections of the prefectures of Macedonia, Thessaly, Thrace, and Epirus. The climate is classified as transitional between that of Mediterranean and Central European. The average air temperature ranges between 11 and 15.4 °C, and the annual

N. Yassoglou et al., *The Soils of Greece*, World Soils Book Series,
DOI 10.1007/978-3-319-53334-6_2

temperature variation is greater than 20 °C. The minimum winter temperatures drop in this region from −2 to −20 °C. The average rainfall ranges between 857 and 506 mm. The distribution of rainfall through the year is more even than in the other climatic regions of the country especially at high elevations. Warm and dry summers prevail in the lowlands.

The soil moisture regime is udic at higher elevations and xeric in the lowlands. The dominant climax soil horizons at elevations higher the 30 m from the sea level or the inland flood plains is the argillic, whereas in the recent alluvial plains no genetic horizon is present in the soil profile other than a weakly developed cambic horizon in wet sites.

(c) The Ionian or Marine Mediterranean Region

The mild marine Mediterranean is the characteristic climate of this region which extends along the western coastal areas from the border to Albania all the way to the southern tip of Peloponnese. It also includes the Ionian island and the Dodecanese islands in the southern eastern part of the Aegean Sea. The area of this region is 22,442 km^2.

The climate is strongly affected by the influence of the Mediterranean Sea. The winter is mild because it is protected from the northern cold winds by the mountain ranges of the mainland. The average yearly temperature variation is 15.5–17 °C, whereas the mean temperatures range from 17.1 to 18.7 °C. Winter frost days are few and occur mainly in northern coastal zone.

The dominant soil moisture regime is xeric in the hilly and low land and udic at high elevations. The dominant diagnostic horizon at elevation higher than 30 m from the sea level or the inland flood plain is the argillic. Soils in recent alluvial plains lack genetic horizons except weakly developed cambic horizons in wet soils as described above. Imperfectly and poorly drained soils form in the flood plains especially in fine-textured profiles.

(d) The Aegean or Inland Mediterranean Region

This region includes the eastern coastal lands and the islands and the northern section of Crete. The two northern Aegean islands Thasos and Samothraki belong to the northern continental region. This region occupies an area of 45,430 km^2.

The mean annual temperature ranges between 16 and 18.5 °C. The rainfall shows significant geographical variations. The Cyclades Islands and southern Attica, including Athens, have mean annual rainfall ranging between 400 and 655 mm, whereas in the northern suburbs of Athens, northwestern Attica, and the eastern Aegean islands, this range is 660–935 mm.

The soil moisture regime is xeric throughout the region. The dominant diagnostic horizon at elevation higher than 30 m from the sea level or the inland flood plain is the argillic. In several cases, it has rhodic characteristics. The ochric surface epipedon dominates in this region. However, the original climax Epipedon on marly tertiary and quaternary deposits was the mollic. This Epipedon has been removed from the soil profile by erosion and extensive cultivation. Soils in recent plains lack genetic horizons.

(e) The Cretean or Desert-like Mediterranean Region

This region according to Mariolopoulos (1936, 1960) has a climate transitional between that of Mediterranean and desert. This classification was based on mean annual rainfall of 202.1 mm, recorded during the period of 1894–1929 (Settas 1975). However, Karapiperis (1955) disputes the existence of this region on the basis of mean annual rainfall that ranges between 504.9 and 517.2 mm. Anyway, this region is considered as the driest in the country. It should also include the southeastern section of Attica and the islands of the Saronikos Bay.

2.1.1 The Bioclimate of Greece

The bioclimate is a compound climatic index, which is related to the performance and the composition of the flora and the fauna on particular landscape. Therefore, it is a parameter which prescribes the effects both of the climate and of the organisms on a particular soil landscape.

Mavromatis (1980) applied two indices to map the bioclimate of Greece

(i) The pluviothermic quotient of Emberger (1955):

$$Q_2 = \frac{1000P}{\frac{M+m}{2}(M-m)}$$

where P is the mean annual precipitation in mm, M the mean maximum temperature of the hottest month in absolute degrees, and m the mean minimum temperature of the coldest month in absolute degrees. The bioclimatic levels were drawn on a diagram m that was plotted versus Q_2, proposed by Emberger–Sauvage (Sauvage 1963).

On the basis of this diagram, Greece is subdivided into three bioclimatic levels: semiarid, subhumid, and humid. Each level was subdivided into three sections on the basis of the minimum winter temperature m as follows: warm $m > 7$ °C, mild 3 °C $< m <$ 7 °C, and cold 0 °C $< m <$ 3 °C.

(ii) The distinction of the characters of the Mediterranean climate:
   (a) The precipitation–temperature diagrams of Bagnouls–Gaussen (1953) for the determination of the duration and the intensity of the dry period, which begins and ends when $P = 2T$ where: P is the mean monthly precipitation in mm and T the mean monthly temperature 0 °C. Dry months are considered those with $P \leq 2T$.
   (b) The xerothermic index X of UNESCO-FAO (1963), calculated for every month, is as follows:

$$X_m = \left\{ J_m - \left(J_p + \left(\frac{J_{r,b}}{2} \right) \right) \right\} f_h$$

where X_m is the monthly xerothermic index; J_m the days of the month; J_p the rainy days of the month; $J_{r,b}$ the days of dew or fog of the month; and f_h the relative humidity coefficient of the month. For relative humidity $H\%$:

$$40\% < H < 60\%, f_h = 0.9;\ 60\% < H < 80\%, f_h = 0.8;\ 80\% < H < 90\%, f_h + 0.7; H > 90\% f_h = 0.6$$

Mavromatis (1980) prepared the bioclimatic map of Greece at a scale 1:1,000,000 by combining the estimates of the two methods in defining the following mapping units:

The Emberger–Sauvage method estimates the vertical succession of the bioclimate considering the effect of elevation on the temperature and the rainfall, which affect the vertical distribution of the plant societies. The Bagnouls–Gaussen method estimates the duration and the intensity of the drought, which affect the biological performance of the plants.

2.1.2 Climate and Vegetation Affected Soil Properties

The genetic trends of the soils are affected by the characteristics of the respective bioclimatic zone.

Soils in the xero-thermo-Mediterranean zone have been developed under mostly a xerophytic pine and shrub natural vegetation. They have a xeric moisture regime, and they become highly vulnerable to soil erosion and desertification upon losing their difficult-to-recover plant cover. Residual soils, upon long periods of undisturbed conditions, develop chromic and rhodochromic argillic horizons, except on marly deposits, where they develop cambic horizons and mollic epipedons. Due to the extensive erosion and cultivation, the present dominant soil typological units are those of Xerorthents on severely eroded landscapes and Xerochrepts on the less eroded ones. Organic matter content is generally low. On alluvial plains, the dominant soil is the Xerofluvents. Development of soil salinity takes place in some coastal or inland depression landscapes.

Xero-thermo-Mediterranean	$150 < X < 200$
Thermo-Mediterranean	$100 < X < 150$
(a) Intense	$125 < X < 150$
(b) Weak	$100 < X < 125$
Meso-Mediterranean	$40 < X < 100$
(a) Intense	$75 < X < 100$
(b) Weak	$40 < X < 75$
Sub-Mediterranean	$0 < X < 40$
Axeric	$X = 0$
(a) Subaxeric with a period when	$3t > P > 2T$
(b) Temperate axeric	

Genetic trends in the thermo-Mediterranean zone are similar to those of the above zone. However, the effects of removal of the protective plant cover on the soil profile in terms of erosion and desertification are not as acute as in the previous zone. The presence of argillic is still rare, but the cambic horizon is more frequently found. Calcic and petrocalcic horizons are developed on calcareous materials.

Most of the residual soil profiles in both zones are erosional descendants of original Xeralfs and on highly calcareous tertiary and quaternary deposits of original Xerolls.

The meso-Mediterranean zone, under natural conditions, has been dominated by the broad-leaved forests, mainly various species of oak. The dominant soil moisture regime in the intense subzone is the xeric, and the dominant soil horizon is the well-developed argillic. It seems that under oaks and erica, clay illuviation is a favored soil genesis process (Yassoglou 1973; Dafis 1966). According to the second author, clay illuviation is more pronounced under Quercus conferta, an oak tree growing in Greece at middle elevations. At these elevations, rainfall may contribute to clay illuviation and not to extensive leaching and extreme acidification of the profile that could inhibit clay translocation. It is worthwhile the relationships among climatic parameters and the effects of the soil organic matter produced by the litter under oaks and erica be further investigated. The main colors of this horizon are the chromic on almost all parent materials except on limestone or on schists with limestone intrusions, where the argillic horizon develops a rhodochromic hue. Thus, dominant original profile soil typological units are the Haploxeralf and the Rhodoxeralf. A special case is the upper Pleistocene terraces where the soil profiles develop profiles with ultic morphology and base saturation.

In the weak meso-Mediterranean zone prevail the oak forests and both the xeric and the udic soil moisture regimes The kind of these regimes depends on the topography, the elevation, and the aspect of the particular soil landscape. Xeric regimes dominate on lower elevations and southern aspects, where udic is more common on the higher elevation northern slopes. The argillic horizons are well developed and characterize Haploxeralf profiles with the exception of those on limestone which classify the respective profiles as Rhodoxeralfs.

The terrain in both of the bioclimatic subzones, especially in the weak, is sloping, subjecting the soils to significant erosion risk upon removal of the protecting vegetative cover. Large sections of these zones have soil profiles, which have lost parts or all of their genetic horizons. Thus, many of the soils of this zone have been degraded to erosional descendants of the original Alfisols and are classified as Orthents (Regosols) or Orthepts.

The sub-Mediterranean zone includes the mountainous lands at elevations higher than the previous zone. It is a transitional zone between the Mediterranean and the Central European. Dominant species are the broad-leaved oaks, the Cephalonian fir, black pine, and beech. The soil moisture regime is udic, the dominant soil horizon is the argillic, and the typological unit is the Udalf. In the higher elevation sections of the zone, excessive leaching of soil bases and acidification of the profiles, on non-limestone parent materials, block clay illuviation. The dominant genetic diagnostic horizon in these areas is the cambic, and the respective typological unit is the Dystrochrept.

The axeric zone occupies the summits of the high mountains of northern Greece. The climatic conditions are more like Central European than Mediterranean. It is dominated by fir trees, Scotch pine, beech, and spruce. The dominant genetic soil the cambic is the dominant soil horizon and the Dystrochrept the respective typological unit On limestone, however, the argillic diagnostic horizon and the Udalf typological unit develop under forest vegetation, whereas the Argiudoll unit predominates in meadows. On the southern limestone slopes, the argillic horizon has a rhodic color.

The subaxeric bioclimatic zone extends in the summits of the high mountains of the central and southern mainland and Crete. The fir and pine species are a little more xerophytic to adjust a drier month during the summer. The soil profiles are similar to these of the previous zone. However, in the southern mainland and Crete, their argillic horizons and Udalfs are also present.

Soil organic matter content in the mountainous regions increases by a factor of 2 with a decrease of 100 °C in the mean annual temperature along with increasing elevation (Nakos 1984). The same author has found that exchangeable Ca and Mg show a decreasing trend from the Mediterranean to the sub-Mediterranean bioclimatic zone, whereas the total N and extractable P, K, Fe, Mn, and Zn increase.

2.2 Parent Material

The parent materials of the soils of Greece have a profound influence on their morphological, chemical, physical, and mineralogical properties. The age of the profile and the climatic conditions have not been able to wipe out this influence. The soil profile properties, which are strongly dependent on the parent materials, are as follows: the genetic diagnostic horizon, the textural and structural configuration, the pH, the mineralogy, the color, and micromorphology. This is the reason that the distinction of soils of Greece based on soil parent material was preferred to any other way (Fig. 2.1).

Greece belongs to the active Alpine orogenic territory (251 million years ago till today), which is the cause of its complex geological landscape. Its geological formations can be grouped into three major categories (Moudrakis 1985):

(i) Pre-alpine formations. They include the bulk of the metamorphic rocks: gneisses, mica and chlorite schists, marbles, and metamorphosed basic and ultrabasic rocks. Granitic and volcanic intrusions occur in faults formed during the alpine orogenesis.
(ii) Alpine formations. They include limestone of various ages, flysch, and chert. Flysch is extensive in western Greece.
(iii) Post- alpine formations. They are mainly sedimentary formations, which consist of river and coastal conglomerates, alluvial deposits sand stones siltstones, marls of lacustrine or marine origin. These formations are younger than those of the Miocene age, mostly unconsolidated and of low degree of diagenesis. They usually are parent materials of relatively high-quality soils.

The main parent rock formations are (Fig. 2.2):

Acid igneous rocks: Granites, Granodiorites (Pareo-Mesozoic and Tertiary, before 245 million years ago, 245–65 million years ago, and 65 million years ago to date).

Basic and ultrabasic igneous rocks: Gabbros, Diabase, Peridotites, Serpentines, Pyroxenites, Ophiolites.

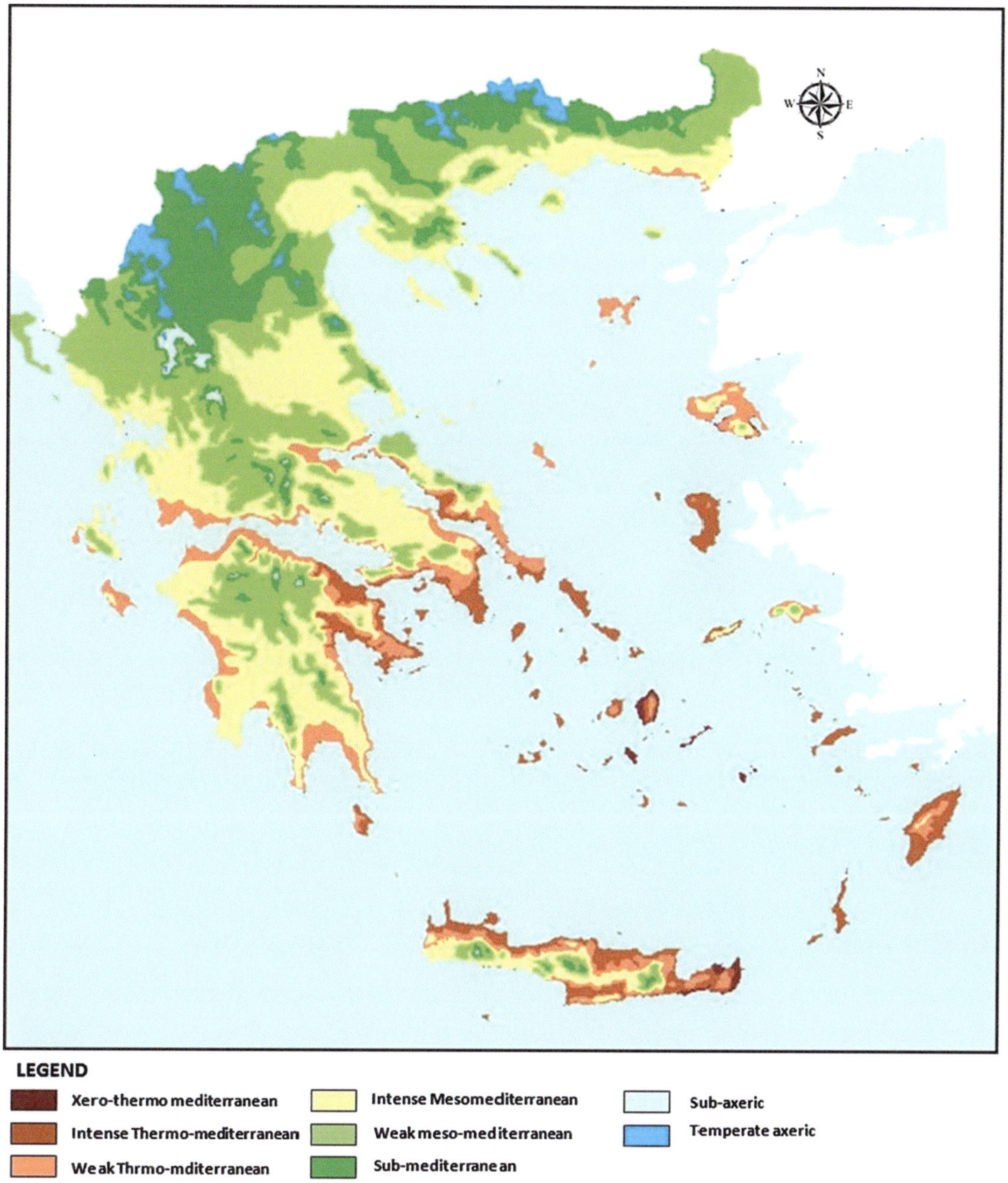

Fig. 2.1 Bioclimatic map of Greece (Mavromatis 1980)

Metamorphic rocks: Gneiss, Mica, Chlorite and Quartz Schists often with layers of Marble, Phyllites (Paleozoic and Triassic, 600 and 208–146 million years ago, respectively). **Flysch**: Interbedded sandstones and shales (Jurassic and Cretaceous, 208–146 and 146–65 million years ago, respectively). **Volcanic rocks:** Rhyolite, Dakite, Andesite, Basalt
Sedimentary Rocks:

- Limestones [Mesozoic (252–66 million years ago), Triassic (251–200 million years ago), Jurassic (200–145 million years ago), Cretaceous (145–66 million years ago).
- Marine and Lacustrine deposits, Marls, Sandstones, Conglomerates, Clays (Oligocene, 38–28 million years ago).
- Marine and Continental Formations (Undivided Quaternary, 1.8 million years ago).
- Marine and Continental deposits, Clays, Loams, Sands, Conglomerates a clayey materials (Pleistocene, 5–1.8 million years ago).

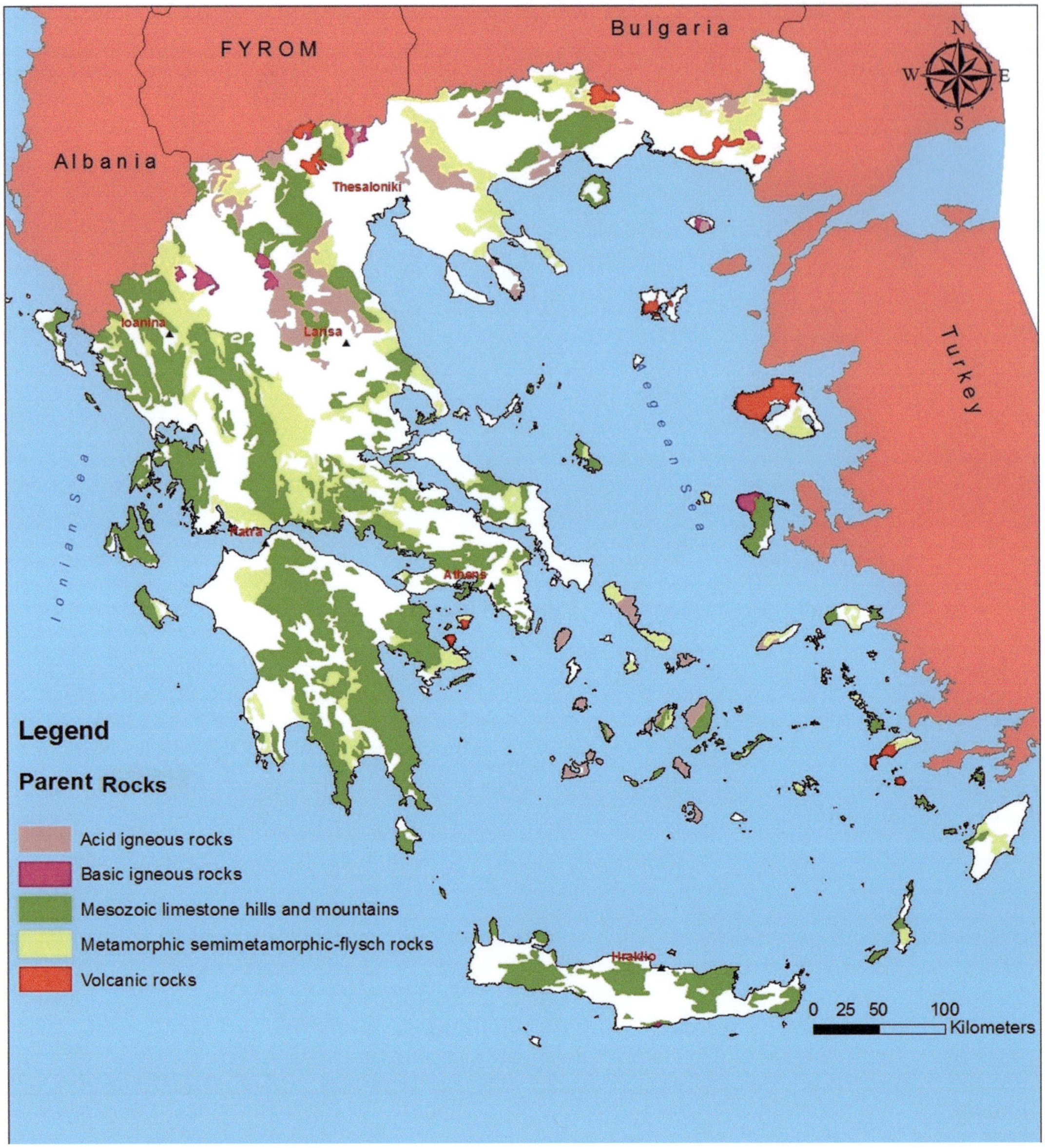

Fig. 2.2 The main parent rock formations in Greece

Alluviums (Holocene, 11,000 years ago)

- Recent Plains
- Older Plains (more than 2000 years old).

2.2.1 Parent Material Affected Soil Properties

Soil properties and morphologies described below refer to undisturbed by human actions or other natural condition profiles. The effects of human action have caused significant deviations in the properties and morphologies of many soils. These deviations will be discussed in the section of the anthropic role on soil genesis.

Soils on acid igneous rocks are usually coarse-textured (loamy sands and sandy loams) and relatively shallow. Their reaction is acid. Their pH on high elevations, where the annual rainfall exceeds 1000 mm, may drop as low as 3.8. Consequently, the base saturation corresponds in most cases to the dystric character of the profile and occasionally to the ultic. At lower elevations, they develop weakly formed argillic horizons.

Acid metamorphic rocks produce soils with a little finer textured (sandy loams and loams) and deeper soils than the acid igneous rocks. Acidity follows the same patterns, except in the case the rocks are interbedded with marble. In the last case, the soil profiles have finer textures, an eutric acidity regime and redder color hues, indicating the formation of hematite.

On basic igneous rocks soils have eutric acidity regime with high base saturation and a mostly neutral pH and fine textures (clay loams to clay). Their profiles form a well-developed chromic argillic horizon, which has frequently vertic characteristics.

Flysch is quite extensive in the mountainous and hilly terrains of Greece. It consists of sandstone, shale, and interbedded layers of both. Genetically related to the soil are the uppermost layers. Sandy layers produce usually sandy loam, loam, and sandy clay loam textures, whereas shale produces finer textures. In cases where the thickness of the layers is smaller than that of the profile, texturally stratified soils are formed. Acidity varies according to the composition of the parent material. In most cases, the pH ranges between 5 and 6 (Nakos 1984). However, in the meso-Mediterranean and axeric zone soils may become strongly acid. Clay illuviation is common in soils with moderate acidity, whereas in the strongly acid soils, no clay illuviation is observed and dystric and ultic cambic horizon is usually formed (Yassoglou 1969, 1973).

Limestone formations have a profound effect not only on the genetic course and properties of the soils but also on the distribution of the vegetation zones disrupting the regular succession with the elevation from the more xerophytic to the less xerophytic. The general characteristics of the residual soils on limestone are as follows:

- A well-structured surface mineral horizon, rich in well-decomposed mull-type organic matter and high biological activity.
- A well-developed argillic horizon, with a clay loam to clay texture and reddish (rhodochromic to chromic) color hues, is formed. In the axeric zone, it may have a brownish hue.
- Soil depth is generally shallow due to the fact that the amounts of soil materials remaining in situ after the dissolution of the carbonates are small.

Marine and lacustrine deposits are calcareous or siliceous. Calcareous sandstones, clays, and conglomerates form soils with pH usually ranging between 7.0 and 8.0. Their textures vary according to the composition of the parent material. Carbonate leaching and formation of calcic and petrocalcic horizons form in their profiles, along with calcaric and eutric cambic horizons. Argillic horizon may form on parent materials not very rich in carbonates such as conglomerates.

Soils developed on marine and lacustrine deposits are usually deep varying in textures according to that of the parent material. In most cases, they range from clay loams to silty clays and clays. They are considered among the productive soils of the country.

Marls, as parent materials, exert a profound influence on the soils, because of the high number of carbonates they contain. Under the Mediterranean climatic conditions, they are difficult to be leached out of the profile. Soils usually maintain residual quantities of carbonates and base over saturation. These conditions prevent clay illuviation and favor strong structural aggregation and formation of mollic epipedons and cambic and calcic horizons.

Sands and conglomerates also contain carbonates in several cases but not so abundant as to prevent clay illuviation.

Pleistocene terraces form chrono-toposequences at elevations usually ranging from 50 to 300 m from the sea level in coastal lands or from the flood plain of the inland areas.

The degree of development increases from the lower terrace to the uppermost in the topo-chronosequence. Their characteristic is the absence of carbonates from the profile, argillic horizon, the increasing acidity, and decreasing base saturation from the lower to the upper terrace. The upper member of the topo-chronosequence develops ultic and paleosolic characteristics. Soils on level terrain in the subhumid areas of the country acquire hydromorphic characteristics and contain abundant Fe–Mn concretion character (Antonakopoulos 1978; Antonakopoulos and Yassoglou 1986). Soils in the drier areas of eastern Greece lack hydromorphic properties. Bisequal profiles have been developed in some of the upper members of the topo-chronosequence.

Soils on Holocene alluviums are the most productive of the country. They can be divided into two groups: (a) those of age older than 2000 years and (b) the recent alluvium of younger than 2000 years.

The first group is geomorphologically located on the Upper Holocene terraces, usually at elevations greater than 20–30 m from the sea level in coastal areas or from inland present flood plains. Soils on these terrains have in many cases formed argillic horizons at variable stage of development. Studies on archaeological sites by Yassoglou and Nobeli (1972) and Haidouti and Yassoglou (1982) have shown that the formation of these horizons in meso-Mediterranean bioclimatic zone takes 2000–3000 years. They could be considered as the youngest argillic horizons found in soils of Greece. The profiles taxonomically belong to the Typic Haploxeralfs (Soil Survey Staff 1975, 1992) or Typic Luvisols (FAO 1998).

The second group consists of recent alluvial deposits at elevations usually less than 20 m from the sea level or inland flood plains. The soils show no or minimum genetic profile differentiation other than the formation of an A1 surface horizon and weak color or texture-differentiated cambic horizons. Weak calcic horizons occasionally form in the imperfectly or poorly drained lower members of the catena.

The soils are in most cases calcareous throughout the profile, stratified and their composition reflects that of the sources of the alluvial sediments. Their textures vary vertically and spatially, according to successively deposited alluvial layers. In many cases, textural and hydromorphic differentiation is observed in relation to the geomorphologic location in the alluvial plain. Soil textures coarser along and close to the river levees, becoming gradually finer away from them. They are also finer and more hydromorphic in the depressions of the flood plain.

An investigation contacted by Kollias et al. (1999) in a holocene alluvial plain of western Peloponnese showed a considerable spatial variation of textural and drainage configuration of the profiles of respective mapping units. The purity of five such units was evaluated using kriging, inverse-squared distance (ISD) interpolation, and conventionally. Soil parameters examined were soil clay content at three depths and the level of the profile drainage Respective fuzzy maps were prepared. The results showed that the overall purity of the five mapping units estimated by kriging, ISD, and conventionally 62.8, 55.7, and 60%, respectively.

Sediments on quaternary terraces and alluvial plains, derived from the erosion of residual soils formed on basic and ultrabasic rocks, develop, on well-drained sites, Vertisols or vertic soil subgroups. These soils usually have fine textures with predominantly expanding clay minerals.

Volcanic rocks are widespread in the Aegean islands and in rather small patches along the northern borders of Greece. Studies conducted by Moustakas and Georgoulias (2005) in the Aegean islands have shown the absence of andic and vitric properties in the soil materials, which are dominated by crystalline aluminosilicates. The only amorphous materials found were the non-crystalline silica. Therefore, the soils even though they have been formed on acid volcanic parent materials (Rhyolites, Dacites, Andesites) cannot be classified as Andisols. They are coarse-textured xeric Entisols and Inceptisols.

In northern Greece, the volcanic rocks are acid Eocene and OligoceneTrahytes and Tuffites which produce fine-textured soils characterized as Vertisols or as Vertic subgroups. Their moisture regimes are udic on high elevations and xeric on low.

2.3 Relief

Relief is also expressed as physiography and topography. A great part of the country consists of rugged mountainous and hilly terrains. Elevation rises within a few kilometers to dozens of kilometers from the sea level to several hundred meters high. Slope gradients, in about 55% of the country, are high enough to cause extensive erosion and significant retrogressive evolution of soil profiles on these lands when they are not adequately protected by vegetation (Table 2.1).

Besides the slope gradient, aspect plays an important role in the erosion risk the moisture regime, the organic matter content and the resilience of the disturbed soils and vegetation, by modifying the local microclimate. Thus, northern, northeastern, and northwestern slopes have a significant wetter microclimate as compared to the southern, southwestern, and southeastern slopes. The erosion risk on the former aspects is significantly smaller on the former than on the latter. Consequently, the depth, the moisture regime, the productivity, and the rate of pedogenetic processes are generally greater on the northern and northeastern slopes and to a lesser degree on the northwestern slopes, as compared to south-oriented slopes.

Topography plays an important role in the hydromorphic characteristics of the soils. Its effect is most pronounced in the more humid western section than in the drier eastern section of the country. Fine-textured soils in the lower parts of the flood plains in the west of the country are usually imperfectly and poorly drained. In most cases, however, the shallow water table is present only during the wet season of the year. In the drier eastern part of the country, poorly drained soils occur only in locations of the alluvial plains where there are permanent shallow water tables supplied by the underground water. Fine-textured soils in these sections of the country are well drained unless when they are supplied by shallow permanent underground water table.

Soil textures in alluvial plains show a significant stratification and usually become finer with distance from the present or past river banks.

Table 2.1 Physiographic distribution of the Greek land (Center of Programming and Economical Studies 1976)

Physiography	Elevation (m)	Area (km^2)	%
Mountainous	>700	56,192	42.6
Semi-mountainous	100–700	35,814	27.1
Level	0–100	39,983	30.3
Total		131,990	100

Slope classes shown in the above map are as follows:

Class	Description	Gradient (%)	Area covered (%)
1	Level	<6	45
2	Slightly sloping	6–18	11.1
3	Steep	18–35	17.8
4	Very steep	>35	26.3
	Other		0.7

2.4 The Biotic Factor

The biotic soil-forming factor can be distinguished into (a) the natural biosystems and (b) anthropogenic influences.

(a) The Natural Systems

The most influential biotic factor is the natural vegetation which along with the soil organisms affects the course and the rate of soil genesis processes. Of course, the vegetation is not entirely independent of the climate and the geology. Thus, its role could be evaluated in situations where the other soil-forming factors are constant. That is in sets of biosequence. Natural vegetation systems are not constant but vary within the time frames of soil genetic changes. Therefore, the present biotic system should be regarded as the present member of a biotic continuum (Fig. 2.3).

The effects of vegetation on soil genesis and properties were discussed above in combination with those of climate. Here is given additional information on the natural vegetation distribution of natural vegetation within the bio-climatic zones of Greece and on its role as soil genesis factor.

Dominant species of forest trees and shrubs with significant effects on soils are distributed as follows (Nakos (1984) and Mavromatis (1980)):

<u>Thermo-Mediterranean Zone:</u>
Pinus halepensis, *Quercus coccifera*, *Pistacia* sp., and *Erica* sp.
<u>Meso- and Sub-Mediterranean Zone:</u>
Quercus conferta, *Quercus pubescens*, *Quercus Cerris*, and *Abies cephalonica*.

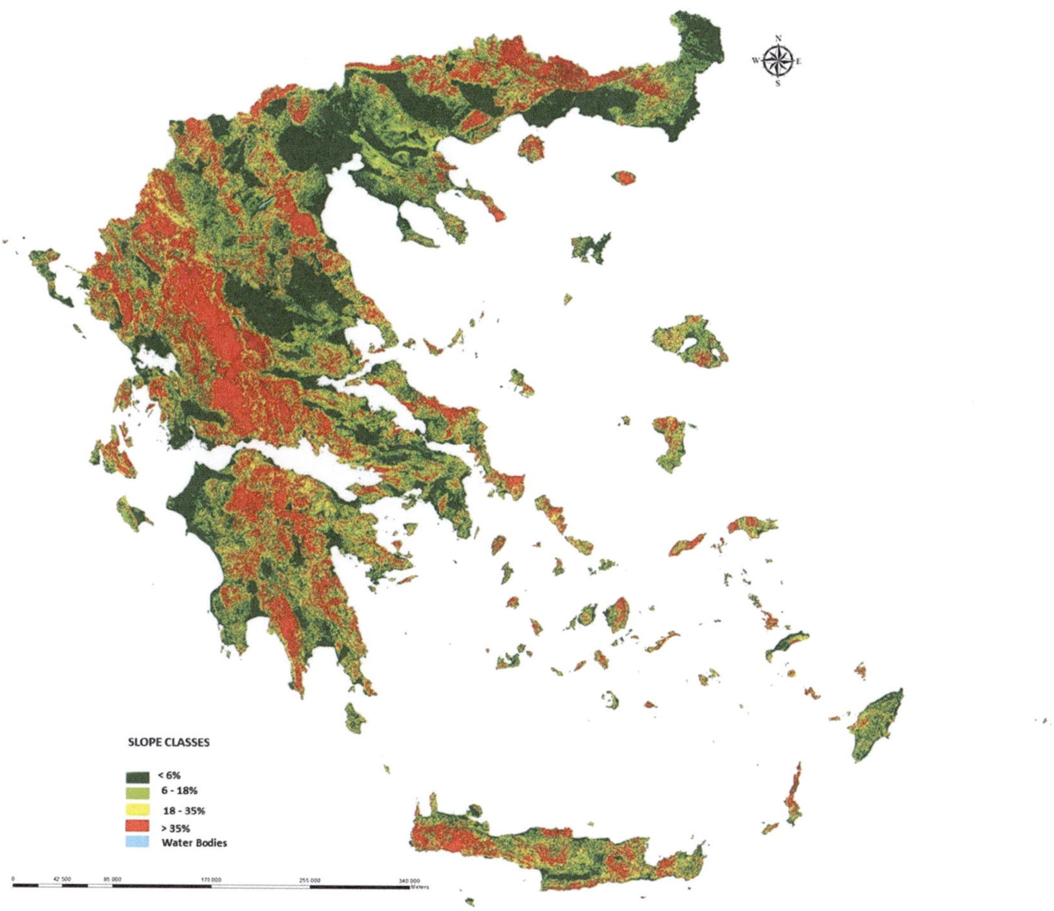

Fig. 2.3 Slope map of Greece (Karamesouti 2011 and Kosmas personal data)

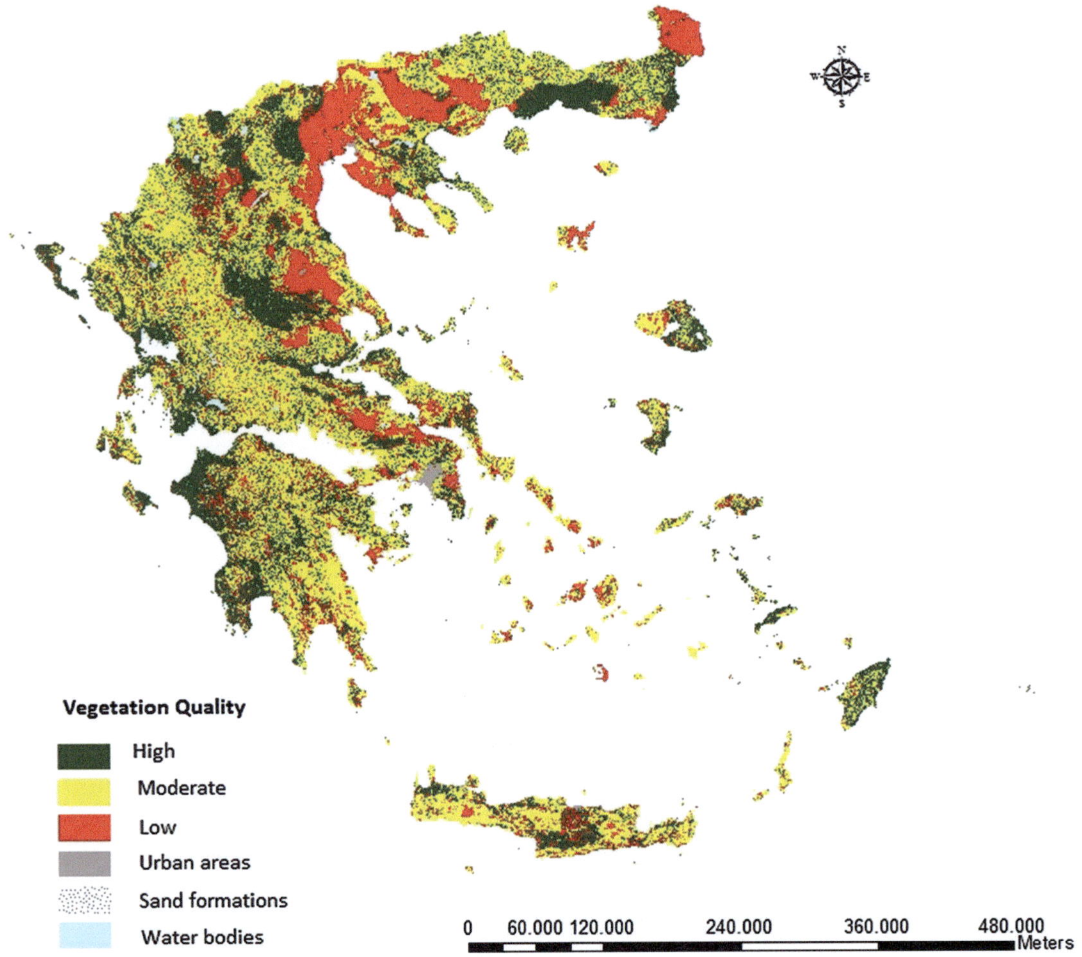

Fig. 2.4 Map of vegetation quality (Kosmas unpublished data; Karamesouti 2011)

Subaxeric and Pseudo-Alpine Zone:
Fagus sp., *Pinus nigra*, *Pinus silvestris*, *Picea excelsa*, and various shrubs and grasses.

The magnitude of the impact of the natural vegetation on soil genesis processes and soil properties depends on the quantity, the composition, and the turnover of its biomass of each particular biosystem.

Kosmas (unpublished data) and Karamesouti (2011) have qualitatively evaluated the role of the vegetation on the Greek terrestrial ecosystems on the basis of its density, resilience, and the degree of protection it offers to the soil against erosion. They have determined vegetation quality, using qualitative indices, and prepared a respective map shown in Fig. 2.4.

The relative areas that the vegetation quality classes in Greece are:

High	Moderate	Low	Other land uses
29.4%	46%	21%	3.6%

The above map and estimates of vegetation quality reflect the present situation. However, the intensity and the effects of human activities on the vegetation and the soil properties varied significantly during the not so remote past.

(b) The Anthropic Factor

Man has effected soil development and properties since about 11,000–13,000 years ago (Arnon 1972). By lumbering, clear-cutting, grazing, and cultivating, land has altered significantly the natural biotic–soil ecosystem. The main effects on soils of Greece were the reduction of organic matter. The deterioration of structure, the decline in surface water infiltration, the truncation of soil profiles through accelerated erosion in sloping terrains, the deposition of fertile sediments in the bottomlands and changes in the chemical and physical properties through cultivation, fertilization, and irrigation. Greeks had employed wise land use systems since at least the Bronze Age. They avoided cultivating marginal sloping land even if it were near their settlements (Yassoglou and Nobeli 1972). Things turned for the worst during the Roman domination. Population and cities grew demanding more food and other supplies. Irrigation and drainage expanded along with river damming. Marginal land brought into cultivation. Inexperienced peasants and slaves caused extensive degradation (Rubio 2007).

Cultivation of undulating hilly lands on tertiary calcareous materials resulted through erosion in the removal or the mollic epipedon and the retrogression of the original soil profiles from Mollisol to Ochrept or event to Orthent.

During the long period of Turkish occupation of Greece, the local Greek population was forced, in many occasions escaping suppression, to move to remote mountainous territories, where unsustainable agriculture, grazing, and forest cutting were practiced in marginal lands.

Over exploitation of forest lands, overgrazing, and unsustainable cultivation of slopping mountainous lands and hill lands on hard rocks resulted in extensive truncation of the original natural soil profile by water erosion. Diagnostic genetic horizons have been removed, and as mentioned in pervious sections, soils retrogressed in large areas from the climax Alfisols Inceptisols and Ultisols to moderately deep and shallow Orthents (Lithic in many cases) and rock outcrops. Undisturbed climax soil profiles are found today only in areas protected by undisturbed natural vegetation.

It should be mentioned, however, that the ruling Turks had in many occasions practiced soil conservation in the more productive lands they held. Agricultural practices have improved the productivity of bottomland soils through cultivation, fertilization drainage and irrigation. In some occasions, however, certain properties such as surface structure, hydromorphic status, salinity, and acidity have been adversely affected through improper practices.

3 Soil Classification

Various soil classification schemes have been applied since the first decades on the twentieth century. Zvorykin Russian soil scientist cooperating with Greek colleagues were the first to apply systematic soil classification and mapping employing the Russian system of soil classification (Nevros and Zvorykin 1937; Zvorikin and Vrisimidji 1947).

Progressive detailed soil survey, employing the American system of soilclassification, was initiated in 1963 by N. Yassoglou and G. Henrard (Second Semiannual Progress Report on Soil Studies in Peloponnese (1963) in Greek).

Land mapping at a scale 1:50,000 has been completed by the Forest Research Institute under the leadership of G. Nakos. It is not a strictly soil survey system, but it contains valuable soil information such as soil profile descriptions and analytical data (Nakos 1983).

Soil laboratories, institutes, and university units have been engaged in various survey projects. The main of them are as follows:

The Soil Institute of Athens, the Soil Institute of Thessaloniki, the N. Kanelopoulos Institute of Athens, the Soil and Agricultural Chemistry Laboratory of the Agricultural University of Athens, the Soil Laboratory of the University of Thessaloniki, the Institute of Soil Mapping and Classification in Larisa, and the Soils Laboratory of the Nuclear Research Center Democritus in Athens.

At present detailed soil surveys are employing Soil Taxonomy and Keys to Soil Taxonomy (Soil Survey Staff 1975, 1992). Soil mapping units (SMU) are designated by soil orders and great groups, whereas soil properties are parametrically designated in the mapping symbol by alphanumeric expressions (Yassoglou 1971). Soil Taxonomy or the World Reference Base for Soil Resources (FAO 1998) are being used for small-scale mapping. The parametric designation of the typological units of these maps was selected in 1964 instead of the descriptive model soil units used in the soil surveys of the USA, Europe, and Russia. There were several reasons for that.

At that time, no efficient detailed model soil mapping units such as soil series had been recognized in the country. This task would have required significant time, effort, and training and would have delayed the urgently needed of a detailed soil map for a scientifically based development of land use and practices in the most productive soils of the country. A disadvantage of the classical soil survey of that time was the model soil series, which were considered as natural bodies that were not accurately defined and recognized in the field. Their usefulness was reduced by the need of introducing variations and inclusions, as well as significant efforts of correlation. A serious advantage of the alphanumeric designation of soil mapping units was anticipated by the foreseen introduction of computers and the electronic handling of data. Furthermore, the system is applied to Quarternary and Holocene terraces, which have a highly stratified profiles, with significant vertical and space variabilities.

The detailed mapping units of soils developed on Quaternary terraces and recent alluvial plains are designated both taxonomically and parametrically. The taxonomic characterization includes symbols of Soil Taxonomy (Soil Survey Staff 1975), referring to soil order, suborder, main group, and subgroup. The alphanumeric symbols refer to:

(a) Soil texture at depths 0–25, 25–75, and 75–150 cm. (b) Landscape or geomorphology. (c) Slope gradient. (d) The presence of abundance and depth of carbonates present in the profile. (e) The degree of drainage of the profile. (f) Limitations that may affect the performance of the soil profile and its productivity.

The above methodology was applied at the initiation of the progressive detailed survey in Greece by Yassoglou and Henrard in 1963, and it was improved by Yassoglou et al. (1983).

N. Yassoglou et al., *The Soils of Greece*, World Soils Book Series,
DOI 10.1007/978-3-319-53334-6_3

An example of the taxonomic-alphanumeric representation of a soil mapping unit (typological unit) in the detailed soil maps is the following:

$$\mathbf{B}\frac{\mathbf{2\,3\,4}}{\mathbf{HA, A, 3, d}}\mathbf{Efxt}$$

B in front of presentation refers to the drainage of the profile (in this case moderately well drained).

The numbers in the nominator correspond as follows: The number **2** corresponds to the texture of the 25–75-cm section of the profile (loamy in this case). The number **3** corresponds to the texture of 75–150-cm section (clayey in this case). The number **4** corresponds to 0–25-cm surface layer of the profile (clay loamy in this case).

The symbols in the denominator correspond the following:

HA to geomorphic landscape (Holocene Alluvium);
A to slope gradient (0–2%);
3 to the presence of carbonates (abundant through the profile);
d to degree of soil erosion assessed on soil surface characteristics; and
Efxt on the right side of the fraction refers to the Soil Taxonomy order, suborder, great group, and subgroup, respectively (Typic Xerofluvents).

Soil associations were used as soil mapping units in the 1:500,000 soil map of Greece (Yassoglou 2004). Soils of Greece were grouped into 45 associations. Each association is designated by the reference soil group, the dominant and associated lower level subdivisions of the reference groups, characterized as typological units. The soil associations, their component typological units (STU), their parent material, and the corresponding Soil Taxonomy taxa are listed in Table 3.1. The geographical distribution of the 45 soil associations are shown in the map of Fig. 3.1. Interpretive attributes to the soil associations are given in Table 3.2.

The principle criterion used in grouping the soil typological units into soil associations was the genetic relation to the dominant climax soil or its retrogressive descendants in each soil environment. Details on this matter are given in this chapter. More information on the properties and functions of the soils are given in Fig. 3.1.

Table 3.1 Soil association map of Greece

Soil order	Soil mapping unit (SMU)	Dominant soil typological unit (STU) FAO	Associated STU FAO	Dominant soil taxonomic unit (Soil Taxonomy)	Associated taxonomic units	Parent material
Lithosols	1	Rock outcrops	Calcaric Lithosol (Ic); Calcaric Leptosol	Rock outcrops	Lithic Xerorthents	Limestone
	2	Calcaric Lithosol (Ic)	Calcaric Regosol (Rc); Calcarochromic Cambisol (Bcc)	Lithic Xerorhtents	Lithic Xerochrepts	
	3	Calcaric Lithosol (Ic)	Calcaric Regosol (Rc); Calcarochromic Cambisol (Bcc); Calcic Kastanozem (Kk); Rhodochromic Luvisol (lcr)	Lithic Xerorthents	Typic Xerorthents; Typic Xerochrepts; Lithic Haploxerolls; Lithic Rhodoxeralfs	Limestone
	4	Eutric Lithosol (Ie)	Eutric Regosol (Re); Vertic Cambisol (Bv); Vertic Luvisol (lv)	Lithic Xerorthents	Typic Xerorthents; Vertic Xerochrepts; Vertic Haploxeralfs	Basic igneous and metamorphic rocks
	5	Eutric Lithosol (le)	Eutric Regosol (Re)	Lithic Xerorthents	Lithic Xerochrepts	Flysch
	6	Eutric Lithosol (Ie)	Eutric Regosol (Re); Eutric Cambisol (Be)	Lithic Xerorthents	Typic Eutrochrepts	
	7	Dystric Lithosol (Id)	Dystric Regosol (Rd); Orthic Acrisol (Ao); Dystric Cambisol (Bd)	Dystric Lithic Udorthents	Dystric Udochrepts; Ultic Dystrochrepts; Dystric Udochrepts	Acid igneous and metamorpic rocks
	8	Eutric Lithosol (Ie);	Dystric Lithosol (Id); Eutric Cambisol (Be); Dystic Cambisol (Bd); Eutric Regosol (Re)	Lithic Udorthents	Dystic-lithic Udorthents; Typic Udorthents; Dystric Udochrepts; Typic Udochrepts	Granites, Diorites, Schists

(continued)

Table 3.1 (continued)

Regosols	9	Calcaric Regosol (Rc)	Calcarochromic Cambisol (Bcc); Calcaric Lithosol (Ic); Rhodochromic Luvisol (Lcr)	Typic Xerorthents	Typic Xerochrept; Lithic Xerorthents; Typic Rhodoxeralfs	Limestone
	10	Eutric Regosol (Re)	Vertic Cambisol (Bv); Eutric Lithosol (Ie); Vertic Luvisol (Lv)	Typic Xerorthents/Typic Udorthents	Typic Xero/Udochrepts; Lithic Xerorthents; Vertic Xer/Udalfs	Basic rocks
	11	Eutric Regosol (Re)	Eutric Cambisol (Be); Eutric Lithosol (Ie)	Typic Xer/Udorthents	Typic Xer/Udochrepts; Lithic Xerorthents	Flysch, Schists, Phyllites
	12	Eutric Regosol (Re)	Eutric Cambisol (Be) Orthic Acrisol (Ao); Chromic Luvisol (lc); Eutric Lithosol (Ie)	Typic Xer/Udorthents	Typic Udochrepts; Typic Haploudults; Typic Hapludalfs; Lithic Udorthents; Typic Udochrepts	Flysch, Schists, Phyllites
	13	Dystric Regosol (Rd)	Dystic Cambisol (Bd); Orthic Acrisol (Ao); Orthic Luvisol (Lo); Eutric Lithosol (Ie)	Dystric Udorthents	Dystric Udochrepts; Lithic Udorthents; Typic Haploudults; Typic Udalfs;	Flysch, Schists, Phyllites
	14	Dystric Regosol (Rd)	Dystric Cambisol (Bd); Dystric Lithosol (Id); Orthic Acrisol (Ao); Orthic Luvisol (Lo)	Dystric Udorthents	Dystric Lithic Udorthents; Typic Haploudults; Typic Udalfs;	Granites, Gneiss Diorite
	15	Eutric Regosol (Re)	Eutric Cambisol (Be); Eutric Lithosol (Ie):Chromic Luvisol (lc); Rhodochromic Luvisol (Lcr)	Typic Xer/Udorthents	Typic Xer/Udochrepts; Typic Xerorthents; Typic Hapludalfs; Typic Rhodoxeralfs	Flysch, Schist, Limestone
	16	Calcaric Regosol (Rc)	Calcic Cambisol (Bc); Vertic Cambisol (Bv); Chromic Vertisol (vc); Calcaric Fluvisol (Jc)	Typic Xerorthents	Calcixerollic Xerochrepts; Vertic Xerochrepts; Chromic Haploxererts; Typic Rhodoxeralfs	Tertiary and Quaternary calcareous deposits
	17	Calcaric Regosol (Rc)	Calcic Cambisol (Bc); Chromic Luvisol (lc); Rhodochromic Luvisol (Lcr)	Typic Xerorthents	Calcixerollic Xerochrepts; Chromic Haploxererts;	Tertiary and Quaternary gravely calcareous deposits
	18	Calcaric Regosol (Rc)	Calcic Cambisol (Bc); Calcaric Fluvisol (Jc)	Typic Xerorthents;	Calcixerollic Xerochrepts; Typic Xerofluvents	Tertiary marls and associated Holocene alluvium
Fluvisols	19	Calcaric Fluvisol (Jc)	Fluvi-calcic Cambisol (Jkf)	Typic Xerofluvents	Fluventic Calcixerollic Xerochrepts; Calcixerollic Xerochrepts	Holocene alluvium
	20	Calcaric Fluvisol (Jc)	Calcic Cambisol (Bc); Rhodochromic Luvisol (Lcr)	Typic Xerofluvents	Calcixerollic Xerochrepts; Typic Rhodoxeralfs	Holocene alluvium, Quaternary terraces
	21	Calcaric Fluvisol (Jc)	Orthic Luvisol (Lo); Eutric Fluvisol (Je) Eutric Cambisol (Be); Chromic Luvisol (lc)	Typic Xerofluvents	Typic Haploxeralfs; Typic Xerochrepts; Typic Haploxeralfs	Holocene alluvium, Holocene terraces
	22	Calcaric Fluvisol (Jc)	Fluvicalcaric Gleysol (Gcf); Gleic Solonchak (Zg)	Typic Xerofluvents	Typic Fluvaquents; Typic Xerofluvents; Aquollic Salorthids	Holocene alluvium
	23	Calcaric Fluvisol (Jc)	Chromic Vertisol (Vc); Pellic Vertisol (Vp); Orthic Luvisol (Lo); Calcic Kastanozem (Kk)	Typic Xerofluvents	Chromic Haploxererts; Typic Haploxerts Typic Haplxeralfs; Calcixerollic Haploxererts	Holocene alluvium, Quaternary terraces
	24	Calcaric Fluvisol (Jc)	Fluvicalcaric Gleysol (Gcf); inclusions of Gleic Solonchak (Zg)	Typic Xerofluvents	Typic Fluvaquents; inclusions of Aquollic Salorthids	Holocene alluvium
	25	Calcaric Fluvisols (Jc)	Eutric Histosols (Oe); Fluvicalcaric Gleisols (Gcf)	Typic Xerofluvents	Sapric Medihemists; Typic Fluvaquents	Lacustrofluvial alluvium

(continued)

Table 3.1 (continued)

Cambisols	26	Calcarochromic Cambisol (Bcc)	Calcaric regosol (Rc); Clacaric Lithosol (Lc)Rhodochromic Luvisol (Lcr)	Typic Xerochrepts	Typic Xerorthents; Lithic Xerorthents; Typic Rhodoxeralfs	Limestone
	27	Calcic Cambisol (Bk)	Calcaric Regosol(Rc); Rhodochromic Luvisol (Lcr); Chromic Luvisol (Lc)	Calcixerollic Xerochrepts	Typic Xerorthents; Typic Rhodoxeralfs; Typic Haploxeralfs	Tertiary unconsolidated deposits
	28	Calcic Cambisol (Bk)	Calcaric Regosol (rc); Rhodochromic Luvisol (Lcr); Chromic Luvisol (Lc)	Calcixerollic Xerochrepts	Typic Xerorthents; Typic Rhodoxeralfs; Typic Haploxeralfs	Calcareous conglomerates
	29	Calcaro-vertic Cambisol (Bvc)	Calcic Cambisol (Bk); Calcarochromic Vertisol (Vcc); Calcic Kastanozem (Kk)	Vertic Xerochrepts	Calcixerollic Xerochrepts; Chromic Haploxererts; Typic Calcixerolls	Tertiary and Quaternary Calcareous terraces
	30	Eutric Cambisol (Be)	Eutric Regosols (Re); Eutric Lithosol (Ie); Chromic Luvisol (Lc)	Typic Xer/Udochrepts	Typic Xer/Udorthents; Lithic Xer/Udorthents; Typic Haploxeralfs	Flysch, Schists, Phyllites
	31	Calcic Cambisol (Bk)	Vertic Cambisol (Bv); Vertic Luvisol (Lv); Eutric Regosol (Re); Eutric Lithosol (Ie)	Calcixerollic Xerochrepts	Vertic Xerochrepts; Vertic Haploxeralfs; Typic Xerorthents; Lithic Xerorthents	Basic igneous rocks
	32	Dystric Cambisol (Bd)	Dystric Regosol (Rd); Orthic Acrisol (Ao); Orthic Luvisol (Lo); Dystric Lithosol (Id)	Dystric Udochrepts	Dystric Udorthents; Typic Hapludults; Typic Hapludalfs; Dystic-lithic Udorthents	Flysch, Gneiss, Schists
	32a	Eutric Cambisol (Be)	Eutric Regosol (Re); Chromic Luvisol (Lc); Eutric Lithosol (Ie) Orthic Andosol (To)	Typic Xer/Udochrepts; Typic Hapolxer/Udalfs	Typic Xer/Udorthents; Typic Xer/Udalfs	Acid volcanic rocks (mainly)
	33	Calcaro-vertic Cambisol (Bv)	Eutric Cambisol (Be); Calcarochromic Vertisol (Vcc); Calcaro-pellic Vertisol (Vcp); Calcaric Fluvisol (Jc)	Vertic Xerochrepts	Typic Xerochrepts; Chromic Xererts; Typic Xererts; Typic Xerofluvents	Holocene alluvium and older terraces
Vertisols	34	Calcarochromic Vertisol (Vcc)	Calcaro-vertic Cambisol (Bvc); Pellic Vertisol (Vp); Calcaric Fluvisol (Jc)	Chromic Xererts	Vertic Xerochrepts; Typic Xererts; Typic Xerofluvents	Eutric Cambisol (Be)
	35	Chromic Vertisol (Vc)	Chromo-vertic Luvisol (Lvc); Calcaric Regosol (Rc)	Chromic Xererts	Vertic Xeralfs; Typic Xerorthents	Tertiary deposits
Luvisols	36	Chromic Luvisol (Lc)	Calcic Cambisol (Bk); Calcaric Regosol (Rc)	Typic Haploxeralfs	Calcixerollic Xerochrepts; Vertic Xeralfs	Unconsolidated Tertiary deposits
	37	Rhodochromic Luvisol (Lcr)	Eutric Cambisol (Be); Eutric Regosol (Re)	Typic Rhodoxeralfs	Typic Xerochrepts; Typic Xerorthents	Quaternary Pleistocene conglomerates
	38	Chromic Luvisol (Lc)	Eutric Planosol (We); Rhodochromic Luvisol (Lcr)	Typic Haploxeralfs	Typic Palexeralfs; Typic Rhodoxeralfs	Tertiary and Quaternary Pleistocene unconsolidated deposits
	39	Chromic Luvisol (Lc)	Eutric Cambisol (Be); Eutric Regosol (Re); Orthic Acrisol (Ao); Eutric Lithosol (Ie)	Typic Haploxeralfs	Typic Xer/Udochrepts; Typic Xer/Udorthents; Typic Udults; Typic Udorthents	Flysch, Schists, Gneiss
	40	Vertic Luvisol (Lv)	Vertic Cambisol (Bv); Eutric Regosol (Re); Eutric Lithosol (Ie)	Vertic Haploxeralfs	Vertic Haploxeralfs; Typic Xerorthents; Lithic Xerorthents	Basic igneous rocks
	41	Orthic Luvisol (Lo)	Orthic Acrisol (Ao); Dystric Cambisol (Bd); Eutric Lithosol (Ie); Dystric Lithosol (Id)	Typic Haploxer-udalfs	Typic Udults; Dystric Udochrepts; Typic Udorthents; Dystric Udorthents	Flysch, Schists, Gneiss, Phyllites
	42	Vertic-calcic Luvisol (Lkv)	Rhodochromic Luvisol (Lcr); Chromic Vertisol (Vc); Calcaric Regosol (Rc)	Vertic-calcic Haploxeralfs	Typic Rhodoxeralfs; Chromic Xererts; Typic Xerorthents	Conglomerates
	43	Orthic Luvisol (Lo)	Eutric Cambisol (Be); Eutric Regosol (Re)			Unconsolidated Quaternary deposits
	44	Chromic Luvisol (Lc)	Eutric Cambisol (Be); Eutric Fluvisol (Je)	Typic Haploxeralfs	Typic Xerochrepts; Typic Xerofluvents	Quaternary terraces, non-calcareous alluvium

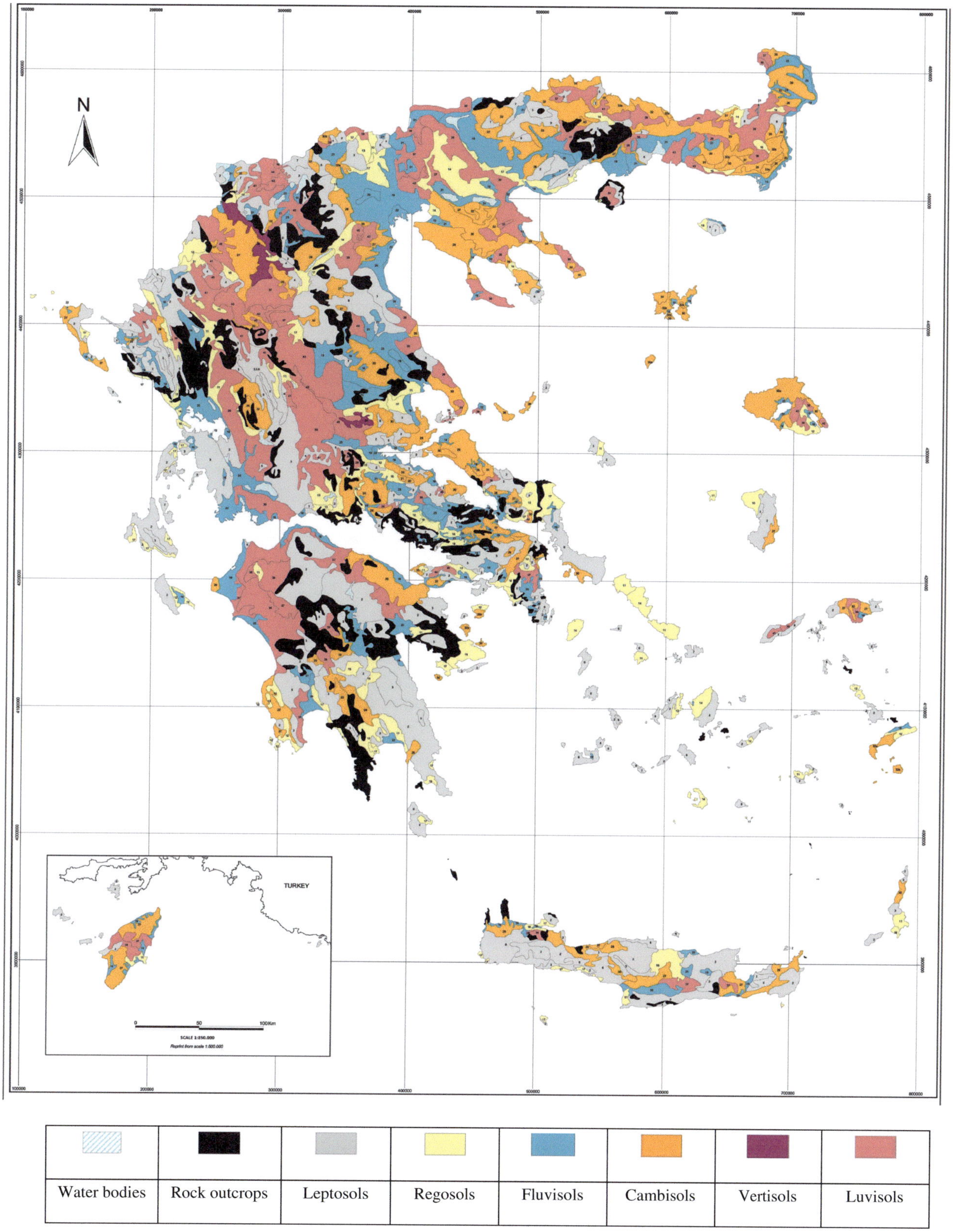

Fig. 3.1 Soil association map of Greece (detailed legend: see Tables 3.1 and 3.2) (Greek National Committee Combating Desertification, Agricultural University of Athens, Prepared by N.J. Yassoglou)

Table 3.2 Attribute table of the mapping units of the 1:500,000 soil association map of Greece

Soil reference group	SMU	Dominant soil typological unit STU[a]	Associated STU[a]	Soil quality	Desertification vulnerability	Sustainable agronomic uses[b]	Restrictions for non-agronomic uses
Rock outcrops	1	Rock outcrops	Calcaric, Eutric, Dystric Leptosols	Very low	Very high	Wild nature	Weak
Leptosols	2	Calcaric Leptosol (PLca)	Calcaro-leptic Regosol; Calcaro-petric Cambisol; Rock outcrops	Low	Very high	Wild nature	Weak
	3	Calcaric Leptosol (LPca)	Calcaro-leptic Regosol; Calcarochromic Cambisol; Calcaro-petric Regosol; Calcic Kastanozem; Rhodochromic Luvisol	Low	Very high	Wild nature, forest, controlled pasture	Moderate
	4	Eutric Leptosol (LPeu)	Lepto-eutric Regosol; Vertic Cambisol; Vertic Luvisol	Low	Very high	Forest, controlled pasture, wild nature	Moderate
	5	Eutric Leptosol (LPeu)	Eutric Regosol; Lepto-eutric Regosol	Low	Very high	Forest, wild nature	Strong
	6	Eutric Leptosol (LPeu)	Lepto-eutric Regosol; Eutric Cambisol; Haplic Phaeozem	Low	Very high	Forest, controlled pasture, wild nature	Strong
	7	Dystric Leptosol (LPeu)	Distro-petric Regosol; Haplic Acrisol; Dystric Cambisol	Low	Very high	Δάσος, ελεγχόμενη βοσκή, άγρια φύση—forest, controlled pasture, wild nature	Μέτριοι – Moderate
	8	Eutric Leptosol (LPeu)	Dystric Leptosol; Haplic Phaeozem; Eutric Cambisol; Dystic Cambisol; Eutro-petric Regosol	low	Very high	Forest, controlled pasture, wild nature	Moderate
Regosols	9	Calcaric Regosol (RGca)	Calcarochromic Cambisol; Calcaric Leptosol Rhodic Luvisol	Low	Very high	Forest, controlled pasture, wild nature	Moderate
	10	Eutric Regosol (RGca)	Vertic Cambisol; Eutric Leptosol; Vertic Luvisol	Low	Very high	Forest, controlled pasture	Moderate
	11	Eutric Regosol (RGeu)	Eutric Cambisol; Eutric Leptosol; Haplic Phaeozem	Low	High	Forest, controlled pasture	Strong
	12	Eutric Regosol (RGeu)	Eutric Cambisol; Haplic Acrisol; Haplic Phaeozem; Chromic Luvisol; Eutric Leptosol	Low	High	Forest, controlled pasture	Strong
	13	Dystric Regosol (RGdy)	Dystic Cambisol; Haplic Acrisol; Orthic Luvisol; Eutric Lithosol	Low	High	Forest, controlled pasture	Strong
	14	Dystric Regosol (RGdy)	Dystric-petric Cambisol; Dystric Leptosol; Haplic Acrisol; Haplic Luvisol	Low	High	Forest, controlled pasture	Moderate
	15	Eutric Regosol (RGeu)	Eutric Cambisol; Eutric Leptosol; Chromic Luvisol; Rhodic Luvisol	Medium	High	Forest, controlled pasture	Strong
	16	Calcaric Regosol (RGca)	Calcaric Cambisol; Haplic Cambisol; Vertic Cambisol; Chromic Vertisol; Calcaric Fluvisol	Medium	Medium	Forest, controlled pasture	Moderate
	17	Calcaric Regosol (RGca)	Calcaric Cambisol; Chromic Luvisol; Rhodic Luvisol	Medium	Medium	Forest, controlled pasture and agriculture	Moderate
	18	Calcaric Regosol (RGca)	Calcaric Cambisol; Haplic Calcisol; Calcaric Fluvisol	Medium	Medium	Forest, controlled pasture and agriculture	Moderate
Fluvisols	19	Calcaric Fluvisol (FLca)	Haplic Calcisol; inclusions of Solonchak in some cases	Very high	Low	Agriculture	Very strong
	20	Calcaric Fluvisol (FLca)	Calcaric Cambisol; Haplic Calcisol; Rhodic Luvisol	Very high	Low	Agriculture	Very strong
	21	Calcaric Fluvisol (FLca)	Haplic Luvisol; Eutric Fluvisol; Eutric Cambisol; Chromic Luvisol	Very high	Low	Agriculture	Very strong
	22	Calcaric Fluvisol (FLca)	Fluvicalcaric Gleysol; Gleic Solonchak	Very high	Low	Agriculture	Very strong
	23	Calcaric Fluvisol (FLca)	Chromic Vertisol; Pellic Vertisol; Haplic Luvisol; Calcic Kastanozem	Very high	Low	Agriculture, pasture	Very strong
	24	Calcaric Fluvisol (FLca)	Fluvicalcaric Gleysol; inclusions of Gleic Solochak	Medium	High	Agriculture	Very strong
	25	Calcaric Fluvisols (FLca)	Eutric Histosols; Fluvicalcaric Gleysols	Very high	Low	Agriculture	Very strong

(continued)

Table 3.2 (continued)

Cambisols	26	Calcarochromic Cambisol (CMcrca)	Calcaric Regosol; Clacaric Leptosol; Rhodic Luvisol	Medium-low	High	Forest, controlled pasture	Strong
	27	Calcaric Cambisol (CMca)	Haplic Calcisol; Calcaric Regosol; Rhodic Luvisol; Chromic Luvisol	High	Medium	Forest, controlled pasture and agriculture	Moderate
	28	Calcaric Cambisol (CMca)	Haplic Calcisol; Calcaric Regosol; Rhodic Luvisol; Chromic Luvisol	Medium	Medium	Forest, controlled pasture	Moderate
	29	Calcaro-vertic Cambisol (CMvtca)	Haplic Calcisol; Calci-chromic Vertisol; Calcic Kastanozem	High–medium	Medium	Forest, controlled pasture and agriculture	Moderate
	30	Eutric Cambisol (CMeu)	Eutric Regosols; Eutric Leptosol; Chromic Luvisol	Medium–low	High	Forest, controlled pasture	Strong
	31	Calcaric Cambisol (CMca)	Haplic Calsisol; Vertic Cambisol; Vertic Luvisol; Eutric Regosol; Eutric Leptosol	Medium–low	Medium	Forest, controlled pasture	Moderate
	32	Dystric Cambisol (CMdy)	Dystric Regosol; Haplic Acrisol; Haplic Luvisol; Eutric Lithosol	Low–medium	High	Δάσος, ελεγχόμενη βοσκή – Forest, controlled pasture	Ισχυροί – Strong
	32A	Eutric Cambisol (CMeu)	Eutric Regosol; Eutric Leptosol; Chromic Luvisol; Haplic Andosol;?	Medium–low	High	Forest, controlled pasture, wild life	Moderate, weak
	33	Calcaro-vertic Cambisol (CMvtca)	Eutric Cambisol; Calci-chromic Vertisol; Pellic Vertisol; Calcaric Fluvisol	High	Low	Agriculture	Very strong
Vertisols	34	Calci-chromic Vertisol (VRcrcc)	Calcaro-vertic Cambisol; Pellic Vertisol; Calcaric Fluvisol	High	Low	Controlled agriculture and pasture	Strong
	35	Chromic Vertisol (VRcr);	Chromo-vertic Luvisol; Calcaric Regosol	High	Medium	Controlled agriculture and pasture	Strong
Luvisols	36	Chromic Luvisol (LVcr)	Haplic Calcisol; Calcaric Regosol	High	Medium	Controlled agriculture and pasture, forest	Moderate
	37	Rhodic Luvisol (LVro)	Eutric Cambisol; Eutric Regosol	High medium	Medium	Controlled pasture and agriculture, forest	Moderate
	38	Chromic Luvisol (LVcr)	Eutric Planosol; Calcaric Fluvisol; Rhodic Luvisol	High	Medium	Controlled agriculture and pasture, forest	Strong
	39	Chromic Luvisol (LVcr)	Eutric Cambisol; Haplic Phaeozem; Eutric Regosol; Orthic Acrisol; Eutric Leptosol	Medium	Medium	Forest, controlled pasture	Strong
	40	Vertic Luvisol (LVvt)	Vertic Cambisol; Eutric Regosol; Eutric Leptosol	Medium	Medium	Forest, controlled pasture	Moderate
	41	Haplic Luvisol (LVha)	Haplic Acrisol; Dystric Cambisol; Eutric Leptosol; Dystric Leptosol (Id); Haplic Phaeozem	Medium	Medium	Forest, controlled pasture	Moderate
	42	Vertic-calcic Luvisol (LVccvt)	Rhodic Luvisol; Chromic Vertisol; Calcaric Regosol	Medium	Medium	Δάσος ελεγχόμενη βοσκή	Moderate
	43	Haplic Luvisol (LVha)	Eutric Cambisol; Haplic Phaeozem; Eutric Regosol; Eutric Fluvisol	High	Medium	Controlled agriculture and pasture	Moderate
	44	Chromic Luvisol (LVcr)	Eutric Cambisol; Eutric Fluvisol	High	Medium	Forest, controlled agriculture and pasture	Moderate

[a]FAO classification, [b]qualitative assessments

4 Residual Soils on Mountains and Hills

Residual soils formed on mountains and hills occupy the 62.6% of the total area of the country. Their classification, morphology, and properties are strongly related to geomorphology, surface geology, and climate as described above. The latter is strongly related to elevation and has a dominant effect on the distribution of the plant communities. The distribution of the soil taxa according to the surface geology (parent material) is shown in Table 4.1. The characteristics, properties, and functions of the soils on this landform developed on each particular parent material are described in the following section. Colluvial surface soil horizons and deep deposits form on some steep mountain and hilly slopes. It should, however, be mentioned that a part of the soil materials is not indigenous, but Aeolian deposits are being transported from Africa as dust particles of fine silt and clay, which have been assimilated by the prevailing pedoenvironment (Yaalon 1997; Mizota et al. 1988).

Aspect and slope, affecting available soil moisture and erosion, have a significant effect on the quality, productivity, ecological functions, and resistance or vulnerability to degradation. Favorable aspects are the northern and the northeastern, and most unfavorable aspects are the southern and southeastern.

The vegetation zones and their characteristic plant species are as follows:

Mediterranean, with the species: *Pinus Halepensis*, *Quercus coccifera*, *Pistacia* sp., *Arbutus* sp., *Erica* sp., *Arbutus* sp., and *Phlomis fluticosa*.

Sub-Mediterranean, with the species: *Erica* sp., *Arbutus* sp., *Pinus Halepensis* and *Pinus brutia*, *Quercus conferta*, *pubescens*, and *cerris*.

Mountainous, with the species: *Abies cephalonica*, *Pinus nigra*, *Fagus* sp., *Pinus silvestris*, *Picea exielsa*, and *Pinus heldreihii*.

Pseudo alpine, with species: *Astrgalus* sp., *Daphne oleoidides*, *Juniperus communis*, *Berberis* sp., *Festuca* sp., and *Stipa* sp.

The basic characteristics of the soils are described in the previous two chapters. In this chapter, they will be discussed in more detail.

It is worth to emphasize here the dominant role of the rootable depth of the residual soils in terms of their biomass productivity, land use suitability, erosion tolerance, hydrology and desertification vulnerability, and resilience. Shallow soils present a critical land use limitation. Soil depth classes in Greece are shown in the map of Fig. 4.1 and Table 4.2 prepared by Kosmas (unpublished data) and Karamesouti (2011).

The data presented in this chapter are values given by the cited authors for soils in various vegetation zones averaged over the zone and represent the mean over the horizons of the profiles studied. They are strongly affected by the basic characteristics of the soils described in the previous two chapters. In this chapter, they will be discussed in more detail. Their classification and the dominant and associated typological units in the respective soil associations are according to FAO (1998) and Soil Taxonomy (Soil Survey Staff 1992), which were listed in Table 3.1 of Chap. 3.

4.1 Soils on Mesozoic Limestones

The dominant climax soil order in all the vegetation zones of Greece is the Alfisol. However, as explained in chapters A and B, climax soil profiles and areas they are found in are limited in number and space on sloping terrains. This is particularly true on limestone formations. There are several reasons for this:

Erosion tolerance is particularly low, and resilience of destroyed or thinned protective vegetation cover is low. This is due to the fact that soil profiles are relatively shallow and the generally high water permeability of the limestone permeability induces conditions of insufficient soil moisture conditions to satisfy the needs for a quick restoration of the damaged vegetative cover.

N. Yassoglou et al., *The Soils of Greece*, World Soils Book Series,
DOI 10.1007/978-3-319-53334-6_4

Table 4.1 Distribution of residual soils into groups of parent material

Parent material	Soil order	MSU*	Dominant STU	Associated STU
Mesozoic limestone hills and mountains		1	Rock Outcrops	Lithic Xerorthents
	Entisol	2	Lithic Xerorthents	Lithic Xerochrepts
	Entisol	3	Lithic Xerorthents	Typic Xerorthents
	Entisol	9	Lithic Xerorthents	Typic Xerochrepts; Typic Xerorthents; Typic Rhodoxeralfs
	Inceptisol	26	Typic Xerochrepts	Typic Xerorthents; Lithic Xerorthents; Typic Rhodoxeralfs
Basic igneous rocks	Entisol	4	Lithic Xerorthents	Typic Xerorthents; Vertic Xerochrepts; Vertic Haploxeralfs
	Entisol	10	Typic Xero/Udorthents	Typic Xero/Udochrepts; Lithic Xerorthents; Vertic Xero/Udalf
Acid igneous rocks	Entisol	7	Dystric Lithic Udorthents	Typic Dystrochrepts; Ultic Dystrochrepts
	Entisol	8	Lithic Udorthents	Dystic Lithic Udorthents; Typic Dystrochrepts
	Entisol	14	Dystric Udorthents	Dystric Lithic Udorthents; Typic Hapludults; Typic Udalfs
Metamorphic–Semi-metamorphic–Flysch rocks				
	Entisol	11	Typic Xero/Udorthents	Typic Xero/Udochrept; lithic Xerorthents
	Entisol	12	Typic Xero/Udorthents	Typic Udochrepts; Typic Hapludults; Lithic Udorthents
	Entisol	13	Dystric Udorthents	Typic Udochrepts; Typic Hapludults
	Entisol	15	Kandic Xero/Udorhtents	Typic Xero/Udochrepts; Typic Hapludults; Typic Rhodoxeralfs
	Alfisol	39	Typic Haploxeralf	Typic Xero/Udochrept; Typic Udult; Typic Udorthent
	Alfisol	41	Typic Haploxeralfs	Typic Udults; Dystric Udochrepts; Dystric Udorthents
Volcanic soils	Cambisol	32A	Typic Xero/Udochrepts	Xero/Udorthents; Xero/Udalfs; Typic Haplo–Xero–Udands

*Soil mapping units of the 1:500,000 soil association map of Greece (Yassoglou 2004)

The climax profiles have the genetic horizons: O-organic top horizon (under forest vegetation), A1-organic matter-enriched mineral surface horizon, Ae difficult to distinguish eluvial horizon, Bt-well-developed illuvial horizon, C-semi-decomposed limestone parent material, and R-undecomposed limestone rock.

The characteristic diagnostic horizon Bt horizon is red colored justifying the designation of the soil profile as Rhodoxeralf (Chromic and Rhodochromic Luvisol). An Argiudoll profile with brownish Bt horizon and a mollic diagnostic epipedon is formed in some cases in the mountainous zone under grass (Yassoglou et al. 1969) Udalf profiles with a brownish Bt horizon may also found in wet locations of this zone.

The majority of the soils in this landform are sequential erosional descendants of the climax soils with truncated profiles:

Inceptisols (Cambisol) having lost the section of the profile overlying much of the argillic Bt horizon, with the subgroups: Typic Xerochrept, Lithic Xerochrept, Typic Udochrept, and Lithic Udochrept

Entisols (Regosol) having lost all diagnostic soil horizons: The soil profile consists of the C horizon overlying the limestone rock. In some cases, a thin organic matter-enriched A1 horizon overlies the C horizon. The common subgroups are as follows: Typic Xerorthents, Lithic Xerorthents, and Typic Udorthents.

Rock out crops. This formation consists of the terminal stage of soil erosion, where the hard limestone bedrock has surfaced on large portion of a territory. Even though this land is considered desertified, there exists several locations with so-called *tube soils*. Karstic vertical surface fissure cracks and tubes are filled with soil material accumulated in them during erosion processes. In some cases, these vertical soil accumulations are deep enough to develop genetic horizons and support grass, shrubs, and even trees.

4.1.1 Basic Soil Properties

The properties of the soils of this landform, as given by Nakos (1984), Sotiriou et al. (1993), and Economou et al. (1993), vary significantly in the various vegetation zones, affected by the vegetation, climate, and parent material. The data in Tables 4.3 and 4.4 represent undisturbed profiles.

The clay distribution in the soil profiles of all vegetation zones shows a maximum in the B horizon, indicating the formation of the illuvial Bt horizon throughout the climatic zones of Greece. The pH values fall within the slightly acid to neutral range suggesting that the carbonates resist the acidification of the soil profile in all the vegetation–climatic zones.

Organic matter and the C/N ration show an accumulation of partly decomposed leaf litter on the O horizon under forest vegetation and incorporation of a favorable amount of

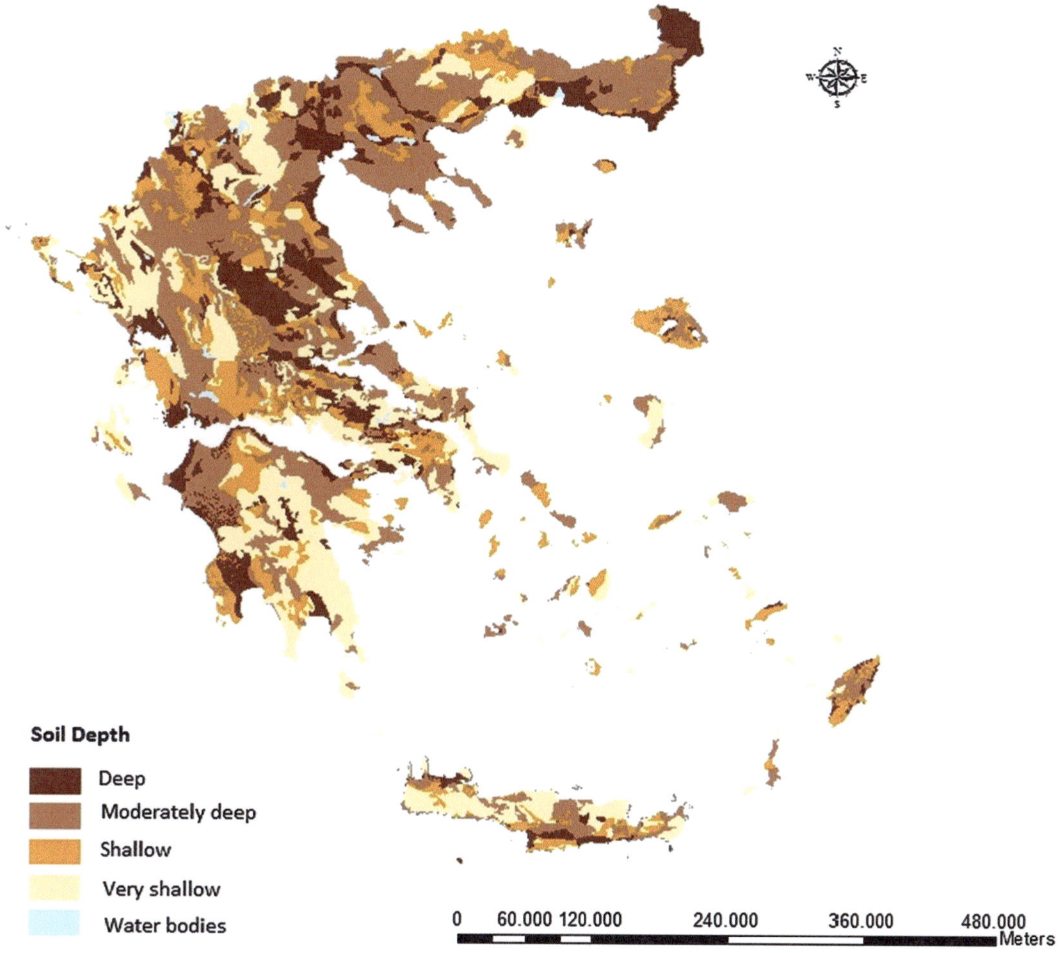

Fig. 4.1 Soil depth classes of soils of Greece (Karamesouti 2011)

Table 4.2 Soil depth classes of soils of Greece (Karamesouti 2011)

Class total area	Description	Depth (cm)	% of
1	Deep	>75	12.7
2	Moderate deep	30–75	32.6
3	Shallow	15–30	21
4	Very shallow	<15	33
Other areas			0.7

well-decomposed organic matter into the surface Ah horizon. These conditions support a high satisfactory biological activity, which contributes to the development of a strong favorable structure on the surface of the soil. This Ah horizon, which takes a relatively short time to form, is also frequently found at various thicknesses in the erosional descendant profiles.

The morphological and analytical data presented in Tables 4.5 and 4.6 are the characteristic of an untruncated climax Rhodoxeralf (Rhodochromic Luvisols) and a severely eroded Lithic Xerorthents (Leptosols) soil profiles.

Designations of soil horizonsand other layers have been adjusted according to the Soil Survey Manual (Soil Survey Division Staff 1993).

Profile: Rhodochromic Luvisol (Rhodudalf)
Region: East Macedonia, northern Greece.
Parent material: Limestone (marble)
Physiography: Mountain slope 20%. Aspect SW
Elevation: 1200
Vegetation zone: Mountainous
Land use: Beech forest and deciduous shrubs

Profile:Calcaric Leptosol (Lithic Xerorthents)
Region: Central Macedonia, northern Greece.
Parent material: Limestone (marble)
Physiography: Steep mountain slope
Elevation: 700 m
Vegetation zone: Sub-Mediterranean
Land use: Deciduous shrubs.

Table 4.3 Ranges of mean values of clay content and chemical properties of soil horizons in profiles of mountainous and hilly soils on limestone

Horizon	Clay (%)	pH	OM (%)	N (%)	C/N	Total P (ppm)	Extr. P (ppm)
O		6.0–6.8	31.3–42.6	0.94–1.1	23–35	700–1036	31.1–40.2
A	20–45	5.8–7.3	3.3–10.7	0.28–0.46	11–17	652–1410	10.4–20.0
B	37–57	5.6–7.3	2.2–5.0	0.14–022	9–13	551–1208	7.8–13
C	17	5.8–7.4	0.7–4.9	0.053–0.21	8–11	410–762	8.6–11.8

Derived from Nakos (1984), Sotiriou et al. (1993), and Economou et al. (1993)

Table 4.4 Ranges of mean values of base saturation, exchangeable cations, and ADTA extractable Fe, Mn, Zn, and Cu from surface horizons of profiles on limestone

Horizon	BS (%)*	Exchangeable cations meq/100 g				EDTA extractable elements in ppm			
		Ca++**	Mg++	K+	Al + 3	Fe	Mn	Zn	Cu
O	52–58	55–56	3.6–6.7	0.2–1.7	1.7	46–89	179–181	30–33	3.6–4.3
A	51–63	25–28	0.7–4.8	0.4–0.7	2.4	30–98	117–131	2.1–2.6	4.3–7.6
B	49–52	29	1.8–10.6	0.3–0.5	2.3	16–70	104–113	0.9–1.3	4.5–6.9
C	61–83	20	0.7–10.8	0.2–0.4	1.2	6–104	2–89	0.8	3.8

Derived from Nakos (1984) and Economou et al. (1993)

*Profiles analyzed are located in the sub-Mediterranean and mountainous zones. **Part of the Ca++ was possibly extracted from the carbonates

Table 4.5 Morphological and analytical data of a Rhodochromic Luvisol (Rhodudalf) profile (Yassoglou et al. 1969)

Horizon	Depth (cm)	Clay (%)	Silt (%)	Sand (%)	Grav. Stones (%)	Structure	Org C (%)	C/N (%)	pH
O	3–0								
A*	0–39	7.1	22.9	70	16.3	mmsbk	0.65	17.0	6.6
2ABt1	39–57	19.9	16.5	63.6	8.4	mmsbk	0.60	12.0	6.5
2Bt2	57–127	29.3	12.5	58.2	7.5	smsbk	0.25	5.0	6.5
2Bt3	127–200	23.8	18.6	57.6	7.8	mmsbk	0.25	5.0	7.0
R	+200								

Horizon	Color Munsell	Exchangeable cations meq/100 g		CEC meq/100 g	BS (%)	Extractable	
		Ca	Ca/Mg			Fe_2O_3 (%)	Al_2O_3 mg/100 g
O							
A	10YR 2/2	5.9	12.5	6.6	86		2.6
2ABt1	5YR 5/6	10.0	10.2	14.2	77	1.09	3.0
2ABt2	2.5YR 5/8	15.6	9.3	19.8	53	0.92	7.0
2Bt3	5YR 4/8	17.7	8.9	19.1	90	0.74	5.8

*The profile has evidently received colluvial and/or alluvial deposits

Table 4.6 Morphological and analytical data of a Calcaric Leptosol (Lithic Xerorthents) profile (Yassoglou et al. 1969)

Horizon	Depth cm	Clay (%)	Silt (%)	Sand (%)	Gravel (%)	Structure	Org.C (%)	Tot N (%)
O	2–0							
A1	0–10	52.0	23.3	34.7	5–15	Crumb	8.9	0.7
A2	10–15	55.0	29.0	16.0	5–15	Crumb	5.6	0.6
R	+15				>80			

Horizon	C/N	$CaCO_3$ (%)	pH	Exchangeable cmol+/kg				CEC, cmol/kg	BS (%)
				Ca	Mg	K	Na		
O									
A1	12.7	45	7.8	>CEC	1.7	0	0.2	69.0	>100
A2	9.3	39	8.0	>CEC	0.8	0	0.2	55.0	>100
R									

Soil fertility estimated from the favorable pH and organic matter in the Ah horizon as well as the availability of plant nutrients: total N, extractable P, Fe, Zn, and Cu, exchangeable K, and Mg, can be considered satisfactory. The carbonates of the limestone bedrock have prevented the acidification of the profile in all vegetation–climate zones enhancing the buffer capacity of the soil.

The key of the productivity and/or quality of the soils of this landform is their depth, which controls the water and nutrient supplies to the plants and the available rooting space. On the base of these criteria, these soils can be listed in the following sequence of declining productivity, land use suitability, ecological contribution and performance, erosion tolerance, resistance to degradation, and resilience from it:

Alfisols > Inceptisols > Entisols > Rock Outcrops.

The presence of a lithic contact in the last two further diminishes their quality and increases their vulnerability to degradation and desertification. A significant fraction of these soils have been already desertified.

The land capability and suitability depend mainly on the depth of the soil profile and the slope gradient on the land. Soils with intact or slightly to moderately eroded Alfisol and Inceptisol profiles can carry productive forest, shrubs pastures. On small slope gradients, they could be suitable for orchards, vegetables, farm crops, and greenhouse cultivation, provided that they are sufficiently supplied with rain and/or irrigation water (Fig. 4.2).

4.2 Soils on Acid Igneous Rocks

Granite and diorite are the main soil parent materials in this landform. The soils are generally coarse with most frequent loamy sand and sandy loam textures. In the mountainous zone, they are usually acid with low levels of base saturation. Due to relatively low rate of weathering of the parent rock, soil depth is usually medium to shallow. Soil acidity and accompanying high levels of extractable Al_2O_3 inhibit the dispersion and illuviation of clay particles (Yassoglou et al. (1969), Pelisek (1964), Ponomayova and Nikolayeva (1964)). Consequently, soil profiles do not form illuvialargillic horizons, except in low elevation under a xeric moisture regime. Thus, the commondiagnostic horizon is the cambic. Thus, the respective climax profile would be a Dystric Udochrept (Dystric Cambisol) and occasionally, under, extreme acid conditions, a Hapludults (Haplic Acrisol). At high elevations under undisturbed vegetation cover, an Umbric epipedon develops on the surface of soil profile. Because of shallowness, the erodibility of their materials and extensive man caused degradation of vegetation cover, and the soils on the steep slopes of this landform are particularly sensitive to erosion. Therefore, a large portion of them have suffered a significant truncation of their profiles. Thus, soils with climax profiles are rare. The majority of them are erosional descendants such as Udorthents with lithic contacts (Leptosols). Rock outcrops occupy extremely eroded and desertified landscapes, but not as many as in limestone formations.

Coarse texture, acidity, and relatively small depth are the characteristic limitations of soils of this landform, inducing conditions of medium-to-low productivity, which decreases with deceasing soil depth. Land use suitability, in most cases, is limited to forestry and pasture. Inceptisols retaining much of their cambic Bw horizon are most productive as compared to shallow Entisols and Ultisols. The presence of lithic contact in the profiles further degrades their productivity (Figs. 4.3 and 4.4).

4.2.1 General Soil Properties

The relatively low rate of weathering of the parent material and the properties of the soils in this landform are the reason for the relatively small depth even in the undisturbed profiles. In the mountainous zone, soil pH and base saturation are generally low, particularly at higher elevations because greater rainfalls and extensive man's activities have caused degradation of vegetation, which has resulted in extensive leaching of the limited amounts bases present in the parent material. In the sub-Mediterranean and Mediterranean zones, soil acidification is less intense allowing the development of illuvialargillic horizons in the untruncated profiles.

Ranges of mean values and morphological and analytical property values of particular soil profiles in this landform are given in Tables 4.7 and 4.8.

Profile: Chromic Acrisol (Typic Hapludults) (P3-Fig. 2.5)
Region: East Macedonia, northern Greece.
Parent material; Diorite
Physiography: Moderate mountain slope
Elevation: 1400 m
Vegetation zone: Mountainous
Land use: Pinus Silvestris Forest.

Profile: Umbric Acrisol (Typic Humaquept) (P4-Fig. 2.5)
Region: East Macedonia, northern Greece.
Parent material; Diorite
Physiography: Gentle mountain slope
Elevation: 1600 m
Vegetation zone: Mountainous
Land use: Picea Excielsa Forest (Table 4.9).

Fig. 4.2 A shallow Calcaric Leptosol (Lithic Xerorthents) (*up*) and a tube soil (*down*) profile on limestone located in the island of Evia and in Attica, respectively (P1, P2-Fig. 4.3)

Vegetation and topography seem to affect the properties and the rate of the development of the soils developed on diorite in the mountainous zone. Under spruce (Picea Excielsa) forest, the disintegration of diorite and the formation of fine materials in the solum are greater than under Scotch pine (Pinus Silvestris). However, one has to consider that the succession of the composition of the vegetation on mountainous soils over the past years and the disturbance by man are not exactly known. Therefore, the conclusion reached above refers to a biotic succession continuum that has presently reached to the present composition of the forest (Fig. 4.5).

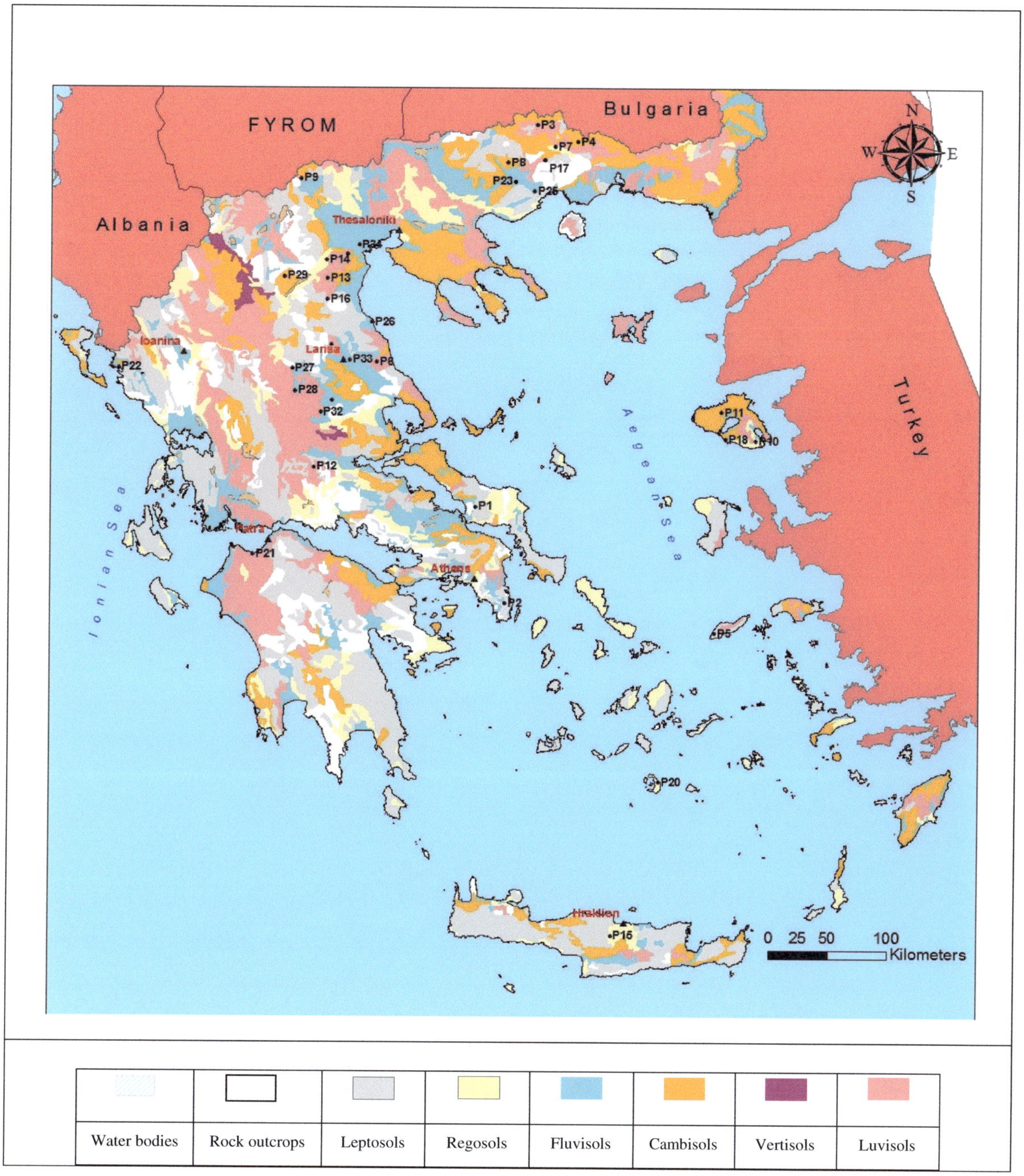

Fig. 4.3 Location map of selected soil profiles. Code number and location of soil profiles. *P1* Calcaric Leptosol (Lithic Xerorthents), island of Evia. *P2* Calcaric Leptosol (Lithic Xerorthents), Attica. *P3* Chromic Acrisol (Typic Hapludults), East Macedonia. *P4* Umbric Acrisol (Typic Humaquepts), East Macedonia Umbric. *P5* Regosol (Lithic Xerorthents), island of Ikaria. *P6* Orthic Acrisol (Udults), Thessaly, central Greece. *P7* Haplic Regosol (Typic Udorthents), East Macedonia. *P8* Eutric Cambisol (Dysrtric Eutorchrepts), East Macedonia. *P9* Dystric Mollic Luvisol (Mollic Hapludalfs), central Macedonia. *P10* Regosol (Lithic Xerorthents), island of Lesvos. *P11* Cambisol (Typic Xerochrepts), island of Lesvos. *P12* Humic Acrisol (Typic Haplohumults), Lamia, central Greece. *P13* Chromic Luvisol (Typic Hapludalfs), Pieria. *P14* Leptic Regosol (Hapludalf Lithic Udorthents), Pieria. *P15* Cambisol (Typic Xerochrepts), island of Crete. *P16* Chromic Luvisol (Typic Hapludalfs), Pieria. *P17* Haplic Luvisol (Typic Hapludalfs), East Macedonia. *P18* Cambisol (Typic Xerochrepts), island of Lesvos—A tube on lava parent material, island of Lesvos. *P20* Arenic Regosol (Vitrandic Xeropsaments), island of Thera. *P21* Eutric Luvisol (Typic Haploxeralfs), northern Peloponnesus. *P22* Calcaric Cambisol (Typic Xerochrepts), Epirus. *P23* Organic soil, East Macedonia. *P24* Calcaric gleyic Fluvisol (Aquic Xerofluvents), Xiniada, central Greece. *P25* Sapric Histosol (Typic Medisaprists), Philippi marshes, East Macedonia. *P26* Calcaric Fluvisol (Typic Xerofluvents), Kouloura, Thessaly, central Greece. *P27* Calcaric Fluvisol (Typic Xerofluvents), Piniada, Trikala. *P28* Calcaric Fluvisol (Typic Xerofluvents), Makrychori, Karditsa. *P29* Calcaric Cambisol (Typic Xerochrepts), Polyfyto, Kozani. *P30* Haplic Luvisol (Typic Palexeralfs), Agia Sofia, Thessaly, central Greece. *P31* Haplic Luvisol (Typic Haploxeralfs), Ryakia, Pieria. *P32* Calcic Vertisol (Typic Calcixererts), Neo Ikonio, Karditsa. *P33* Calcic Vertisol (Typic Calcixererts), Melissohori, Thessaly, central Greece. *P34* Mollic Solonchak (Aquollic Salorthids), Aiginio, Pieria

Fig. 4.4 Distribution of Leptosols in Greece

Table 4.7 Ranges of mean values of clay content and chemical properties of soil horizons in profiles on mountainous acid igneous rocks

Horizon	Clay (%)	pH	OM (%)	N (%)	C/N	CEC me/100 g	Extr. P (ppm)*
A	8.5–18.7	4.4–6.0	3.7–6.8	0.16–0.23	17.3–13	14.4–33.4	5.2
Bw*	13.5	4.7	1.4	0.14	5.9	15.2	
C	3.5–3.5	5.3–6.3	0.26–0.46	0.02–0.04	5.7–13.7	6.6–7.1	0.2

Horizon	BS (%)*	Exchangeable cations meq/100 g			Extractable oxides	
		Ca++	Mg++	K+	Fe_2O_3 (%)	Al_2O_3 mg/100 g*
A	21–60	4.3–6.4	1.9–1.1	0.12	1.1–1.0	41.6
Bw*	19.5	1.4	0.35		0.8	53.5
C	27.8–75	0.8–4.0	0.7–1.4	0.1	0.49–1.0	56.3

Derived from Yassoglou et al. (1969) and Shahabi (1982)
*Mean values

Table 4.8 Ranges of mean values of clay content and chemical properties of soil horizons in profiles on mountainous acid igneous rocks

Horizon	Depth cm	Clay %	Silt %	Sand %	Grav. Stones %	Structure	Org C %	C/N %	pH
O	3–0								
A	0–21	7.4	10.9	81.7	17.9	mmsbk	1.9	19.2	5.2
Bw1	21–27	7.7	11.3	81.0	13.3	wmsbk	1.7	16.8	5.2
Bw2	27–41	9.9	14.2	75.9	12.8	mmsbk	1.5	12.5	5.3
Bw3	41–64	7.1	16.3	76.6	29.9	mmsbk	0.7	13.4	5.2
C	64–140+	2.4	8.5	89.1	8.5	sg	0.2	6.3	5.2

Horizon	Color Munsell	Exchangeable cations meq/100 g		CEC meq/100 g	BS (%)	Extractable	
		Ca	Ca/Mg			Fe_2O_3 (%)	Al_2O_3 mg/100 g
A	10YR3/2	3.0	2.2	19.2	26	0.77	33.1
Bw1	7.5YR4/4	1.9	1.7	16.6	20	0.88	63.0
Bw2	5YR4/4	1.2	1.0	12.5	13	0.84	103.0
Bw3	7.5YR4/6	1.4	1.0	13.4	20	0.71	52.7
C	10YR5/4	–	–	6.3	40	0.47	86.0

Derived from Yassoglou et al. (1969) and Shahabi (1982)

Table 4.9 Morphological and analytical data of an Umbric Acrisol (Typic Humaquept) profile (Yassoglou et al. 1969)

Horizon	Depth cm	Clay (%)	Silt (%)	Sand (%)	Grav. Stones (%)	Structure	Org C (%)	C/N (%)	pH
O	1–0								
A1	0–5	25.3	25.7	49.0	4.3	mmcr	8.3	20.2	3.8
A2	5–10	14.6	19.4	66.0	4.7	mmcr	4.9	16.3	4.0
Bw1	10–35	13.6	17.3	69.1	4.4	mmsbk	2.7	12.7	4.3
Bw2	35–48	11.8	14.2	74.0	14.1	mmsbk	2.3	16.0	4.4
Bw3	48–57	9.0	12.3	78.7	13.9	mmsbk	1.6	13.7	4.8
C1	57–90	3.6	5.9	90.5	41.6	wsbk	1.0	13.0	4.9
C2	90–130+	1.2	3.4	95.4	41.4	single gr	0.2	3.8	5.5

Horizon	Color Munsell	Exchangeable cations meq/100 g		CEC meq/100 g	BS (%)	Extractable	
		Ca	Ca/Mg			Fe_2O_3 (%)	Al_2O_3 mg/100 g
A1	5YR2/1	8.1	3.4	65.5	17	1.68	39.7
A2	7.YR3/2	1.3	1.7	47.0	5.0	1.33	90.3
Bw1	5YR3/3	0.6	1.6	33.8	3.0	1.22	92.1
Bw2	7.5YR2/2	0.6	2.0	25.7	4.0	1.05	76.5
Bw3	7.YR3.4	0.8	1.8	18.1	8.0	0.74	53.9
C1	10YR3/4	0.4	1.8	12.7	6.0	0.84	51.0
C2	10YR4/3	0.3	2.1	4.0	18.0	0.40	17.9

4.3 Soils on Schists

Parent materials of soils grouped in this landform are mica schist, chlorite schists, and gneiss. Soils developed on mica schists are the most extensively investigated in this landform of Greece. Loamy and sandy loamy are the most frequent textures in these soils. The pH, base saturation, and the genetic configuration of the profile are quite variable and largely depend on the influence of carbonates originating from the adjacent marble and limestone. In their presence,

Fig. 4.5 A shallow Lithic Xerorthents profile on acid igneous rock located in the island of Ikaria (P5-Fig. 4.3)

neutral and alkaline with relatively high base saturation profiles are observed. Alfisols (Luvisols) with chromic and rhodochromicargillic horizons are frequently developed on less extensively eroded mountain slopes in the sub-Mediterranean and Mediterranean zones. In the absence of carbonate formations, dystric and ultic acidity levels prevent the illuviation of clay and the formation of argillic horizons. This is particularly true in the mountainous zone. Cambisol and occasionally Ultisols are the characteristic climax soils on mildly eroded slopes.

Like most of the mountain soils of the country, soils of this landform due to their medium texture have high erodibility. Thus, when they loose their protective vegetation they are subjected to greater erosion rates. They profiles have lost, to various degrees, their diagnostic horizons and are reduced to shallow depths. Thus the order: Orthents (Regosols and Leptosols) with Lithic and Leptic characteristics dominate on the eroded sloping lands. Erosion has reduced the soil's rooting space and water storage capacity. These conditions are not favorable for plant growth.

4.3.1 General Soil Properties

The soils of this landform are comparatively more productive than those on the acid igneous rocks because they have profiles with finer textures (loamy, sandy loamy, and sandy clay loamy), higher fertility, and greater water-holding capacity. Their quality and the biomass potential productivity could be considered as moderate because of their favorable texture, moderate rootable depth of the profile, and the sufficient mineral nutrients provided by the bedrock parent material, which is of rather moderate resistance to weathering. They are particularly suitable as forestlands on steeper slopes and as controlled pastures on lower slope open lands.

Their productivity, as forestlands, also depends on the orientation of their schistosity. Veritical or perpendicular to the land's slope orientations, provide greater soil rooting space, greater water infiltration, lower vulnerability to erosion, as well as higher rates of weathering of the parent rock. On the contrary, schistosity parallel to the direction of slope enhances surface water runoff and erosion resulting in less productive shallow soils.

Ranges of average values of soil properties over the various bioclimatic zones of Greece and data of typical profile are given in Tables 4.10 and 4.11:

Profile: Orthic Acrisol (Udults) (P6-Fig. 2.5)
Region: Thessaly, central Greece.
Parent material: Gneiss schist
Physiography: Mountain slope 40–70%. Aspect: S
Elevation: 700 m
Vegetation zone: Sub-Mediterranean
Land use: Oak (Quercus Conferta) forest and deciduous shrubs (Table 4.12).

Profile: HaplicRegosol (Typic Udorthents) (P7-Fig. 2.5)
Region: East Macedonia, Greece.
Parent material: Mica schist

Table 4.10 Ranges of mean values of clay content and chemical properties of soil horizons in profiles of mountainous and hilly soils on mica schists

Horizon	Clay %	pH	OM %	N %	C/N	Total P ppm	Extr. P ppm
O		5.5–6.5	31.7–35.6	1.2–0.85	24–27	637–924	53–32
A	16–27	4.8–6.5	1.3–4.9	0.1–0.24	23–10.7	262–452	9.3–18.5
B	22–31	4.8–6.6	1.2–2.9	0.12–0.08	12–15	251–545	11–7
C	20–24	4.8–6.8	0.5–1.9	0.10–0.08	9–14	219–549	12–8.5

Derived from Papamichos (1979), Nakos (1984), Sotiriou et al. (1993) and Economou et al. (1993), and Shahabi (1982)

Table 4.11 Ranges of mean values of base saturation, exchangeable cations, and EDTA extractable Fe, Mn, Zn, and Cu from surface horizons of profiles on mica schist

Horizon	BS (%)*	Exchangeable cations meq/100 g				EDTA Extractable elements in ppm			
		Ca++**	Mg++	K+	Al + 3	Fe	Mn	Zn	Cu
O	50–90	27–37	3.6–6.5	0.7–1.7		132–335	207–1327	28–31	2.9–3.7
A	16–85	4–18	0.7–5.5	0.2–0.7		42–300	59–92	1.2–2.0	0.9–1.2
B	33–80	0.3–7.5	0.9–5.3	0.2–0.6		25–160	25–165	1.0–1.1	0.9–1.4
C	42–73	04–13.8	3.6–6.5	0.1–0.5		22–105	14–35	1.1–1.5	0.9–1.2

Derived from Nakos (1984); Economou et al. (1993); and Shahabi (1982)

Physiography: Mountain slope 15%. Aspect: NE
Elevation: 500 m
Vegetation zone: Sub-Mediterranean
Land use: Grasses and evergreen oak (Quercus Coccifera) (Table 4.13)

Profile: Eutric Cambisol (Dysrtric Eutorchrepts) (P8-Fig. 2.5**)**
Region: East Macedonia, Greece.
Parent material: Mica schist
Physiography: Mountain slope 4–5%. Aspect: S
Elevation: 700 m
Vegetation zone: Sub-Mediterranean
Land use: Dense oak forest (Table 4.14).

Profile: Dystric Mollic Luvisol (Mollic Hapludalfs) (P9-Fig. 2.5**)**
Region: Central Macedonia, northern Greece.
Parent material: Chlorite schist
Physiography: Steep mountain slope
Elevation: 1500 m
Vegetation zone: Mountainous
Land use: Mountain meadow (Fig. 4.6 and Table 4.15).

Table 4.12 Morphological and analytical data

Horizon	Depth cm	Clay (%)	Silt (%)	Sand (%)	Grav. stones (%)	Structure	Org.mat. (%)	C/N (%)	pH
O	3–0				Rare		25.94		6.8
A*	0–15	17	40	43	Few	wfsbk	2.47	16	4.9
Bt1	15–51	27	38	35	Rare	mmsbk	1.26	17	5.1
Bt2	17–74	25	36	39	common	wfsbk	1.7		5.2
Bt3	74–104	27	36	37	many	wfablk	1.7		5.2
R									

Horizon	Color Munsell	Exchangeable cations meq/100 g			CEC meq/100 g	BS (%)	Extract P (ppm)	Total P (ppm)
		Ca	Mg	K				
O	10YR3/2	13.5	4.25	0.65	33	56	21.6	700
A	7.5YR5/6	5.9	1.8	0.35	14	19	3.2	400
Bt1	2.5YR4/6	10.0	1.83	0.35	14	26		480
Bt2	5YR5/8	15.6	2.33	0.3	15	32		540
Bt3	5YR5.8	17.7	2.5	0.37	18	31		600

By: Anestis and Perris (1986). Ministry of Rural Development and Foods, Greek Forest Service

Table 4.13 Morphological and analytical data

Horizon	Depth cm	Clay (%)	Silt (%)*	Sand (%)**	Grav. Stones (%)	Structure	Org.mat. (%)
A1	0–15	22.5	58.5	19.0	4.8	mmsg	4.0
A2	15–35	25.0	61.0	14.0	4.0	mmsg	2–6
C	35–55	22.0	64.0	14.0	14.0	wmsg	2.4

Horizon	Color Munsell moist	Exchangeable cations meq/100 g			CEC meq/100 g	BS (%)	$CaCO_3$ (%)	Extr. P (ppm)
		Ca	Mg	K				
A1	10YR3/3	32	0.9	0.1	34	98	0.5	1.5
A2	10YR4/2	29	1.1	0.0	34	88	0.0	0.0
C	7.5YR3/4	25	0.8	0.0	29	89	0.0	0.0

By: Shahabi (1981)

Table 4.14 Morphological and analytical data

Horizon	Depth cm	Clay (%)	Silt (%)*	Sand (%)**	Gravel, stones (%)	Structure	Org.mat. (%)	C/N
A1	0–2	21.0	54.0	25.0	0.7	mmcr	11.7	12.6
A2	2–10	21.0	51.0	28.0	5.4	mmcr	5.4	13.7
Bw1	10–25	21.5	52.5	26.0	2.9	mmsgr	2.0	17.0
Bw2	25–38	25.0	52.5	22.5	1.4	mmabk	1.2	16.4
Bw3	38–53	17.5	49.5	33.0	6.9	massive	0.6	18.0
C	53+							

Horizon	Color Munsell moist	Exchangeable cations meq/100 g			CEC meq/100 g	BS (%)	$CaCO_3$ (%)	Extr. P (ppm)
		Ca	Mg	K				
A1	10YR3/2	29.0	4.7	0.8	38.0	91	0	11.0
A2	10YR3/2	19.0	4.4	0.5	28.0	86	0	4.0
Bw1	10YR4/3	18.0	6.5	0.4	28.0	89	0	0.8
BW2	10YR3/3	23.0	7.2	0.2	33.0	93	0	0
Bw3	10YR4/3	18.0	3.1	0.1	27.0	80	0	0

By: Shahabi (1981)
*Silt: 0.02–0.002 mm. **Sand: 0.02–2 mm. ***applied to textural classes in all tables

4.4 Soils on Flysch

Flysch is a tertiary formation consisting of layers' slate, and stone and siltstone at various degrees of diagenesis. As a soil parent material, it is brittle, relatively easily weathered and disintegrated. It produces deep residual soils on the mountain and hilly slopes. However, due to its brittleness, flysch as a parent material itself and the overlying soil have high erodibility. Thus, when exposed unprotected by vegetation on steep slopes are eroded severely and frequently become sources of soil materials transported by surface waters to lowlands as alluvial deposits.

Soil properties, on this rock, are strongly influenced by the composition thickness or proportion of the layers in the parent material. Soils on sand stone usually have sandy loam to sandy clay textures, whereas on slate, sandy clay loam to clay is the dominant textures (Yassoglou 1973; Nakos 1984; Papamichos 1979).

The erodibility of the soils is high especially on sand stones and silt stones. On slates, even though soil erodibility is relatively low, soils are vulnerable to water erosion, because low rates of infiltration produce large surface runoff and limit soil water storage. The latter slows down the recovery of damaged vegetation cover, exposing bare soil surface for long time resulting in extensive furrow and gully erosion.

On the contrary, soil erosion tolerance on soils on flysch is high, because of the significant depth of rootable and water storage space, provided by the soil profile materials and by the brittleness of the parent material.

The climax diagnostic horizons of the soils in this landform are the argillic and the dystric cambic. The argillic

Fig. 4.6 A Regosol (shallow Lithic Xerorthents) (*up*) and a moderate deep terraced (Typic Xerochrept) (*down*) soil profile on schist located in the island of Lesvos (P10, P11-Fig. 4.3)

horizon prevails in Mediterranean and sub-Mediterranean zones at elevation usually less than 1000 m. The dominant climax taxonomic unit would be the soil profiles, the Alfisol (Luvisol) with usually chromic argillic horizon. Occasionally and especially on higher elevations, the argillic horizon may acquire a dystric character due to the leaching bases.

The Dystric Cambic horizon is frequent in the mountainous zone at elevations greater than a 1000 m. Occasionally, the horizon may acquire ultic (acric) character especially in higher elevations due to the excessive leaching of bases. Taxonomically, the undisturbed profiles are classified as Dystrochrepts (Dystric Cambisols, Acrisols, and Phaeozems). Their erosional descendants are Typic or Lithic Udorthens (Typic, Leptic Regosols).

Erosion has degraded a great part of the soils also in this form. The original argillic and the overlying it surface horizons have been removed from the eroded profiles, especially in the Mediterranean and the sub-Mediterranean zones. Thus, the erosional descendants are dominated by Xerochrepts, Usto/Udochrepts, Xerorthents and Udorthents (Eutric or Chromic Cambisols, and Eutric, Leptic, and Haplic Regosols.

The corresponding erosional descendants in the mountainous zone are Typic or Lithic Udorthens (Dystric and LepticRegosols).

Table 4.15 Morphological and analytical data of a Dystric Mollic Luvisol (Mollic Hapludults) profile (Papamichos 1979)

Horizon	Depth (cm)	Clay (%)	Silt (%)	Sand (%)	Grav. Stones (%)	Structure	Org C (%)	C/N (%)	pH
A	0–15	12	44	44	10	Mmcr	7.2	20	5.6
ABt1	15–31	20	44	36	10	Mmcr	3.3	14	5.6
Bt2	31–50	20	36	44	10	Msbk	1.5		5.3
C	50–110+	14	42	44	70	structureless	0.5		5.1

Horizon	Color Munsell	Exchangeable cations meq/100 g			CEC meq/100 g	BS (%)	Extr P (ppm)
		Ca	Mg	K			
A	10YR2/2	10.5	3.7	1.5	22	71.8	7.6
ABt1	10YR2/2	7.5	2.8	1.4	23	51.5	–
Bt2	5YR3/4	4.0	1.4	0.8	18	34.8	–
C	10YR4/4	3.5	1.6	0.3	14	36.9	–

4.4.1 General Soil Properties

Soils in this landform, not being truncated by erosion, have sufficient depth to produce forest and pasture biomass of high productivity. Their productivity, however, declines sharply on sloppy terrains, which have lost their protective vegetation.

Mountain land on flysch had preferably hosted villages and communities of Greek rural population escaping the suppression in the productive lowlands during the centuries of Turkish occupation. Many of these soils had been eroded and suffered decline of their productivity because of overexploitation.

Most suitable for forestry and range management (the latter on gentle slopes) are the soils on mixed thin layers of sandstone, silt stone, and shale with stratification vertical to the slope of the terrain. These soils show a high degree of resilience in case of erosion. On pure shale parent material, water infiltration and permeability as water availability to plants are limited, whereas landslides are frequent. These conditions downgrade the resilience and the quality of the soil–plant systems.

Representative data of the soils in the flysch landform are presented in Tables 4.16, 4.17, and 4.18).

Profile: Humic Acrisol (Typic Haplohumults) (P12-Fig. 2.5)
Region: Lamia, central Greece.
Parent material: Sandstone Flysch
Physiography: Lower part of mountain slope 40–70%.
Aspect: S
Elevation: 1400 m
Vegetation zone: Mountainous
Land use: Fir (Abies Cephalonica forest)

Profile: Chromic Luvisol (Typic Hapludalfs) (P13-Fig. 2.5)
Region: Pieria Region, central Greece.
Parent material: Sandstone Flysch
Physiography: Mountain slope 45%. Aspect: SW
Elevation: 1050 m
Vegetation zone: Mountainous
Land use: Beech (Fagus Silvatica)

Table 4.16 Ranges of mean values of clay content and chemical properties of soil horizons in profiles of mountainous and hilly soils on flysch

Horizon	Clay (%)	pH	OM (%)	N (%)	C/N	CEC* meq/100 g	Total P (ppm)	Extr. P (ppm)
O		6.0–6.4	33–49	0.9–1.2	18–39	100	646–1090	37–59
A	19–35	4.9–5.9	2.5–8.9	0.07–0.33	13–20	33.3	669–743	5.3–9.0
B	17–41	5.0–5.7	0.7–1.8	0.06–0.13	8–14	27.0	184–340	3.8–8.0
C	14–39	5.1–5.8	0.4–1.5	0.04–0.07	6–12	38.7	181–445	4.2–9.4

Horizon	BS (%)*	Exchangeable cations meq/100 g			EDTA Extractable elements in ppm			
		Ca++**	Mg++	K+	Fe	Mn	Zn	Cu
O	57	35–47	4.5–12.3	1.5–2.1	107–131	225–608	21–31	2.3–4.1
A	23–53	5.8–11	1.6–3.7	0.4–0.8	57–133	37–176	0.9–3.7	1.3–2.0
B	15–48	3.1–10.2	1.6–6.6	0.4–0.6	39–170	30–138	0.8–2.5	0.3–3.0
C	14–49	6.4–11.8	2.4–8.2	0.3–0.5	35–118	32–60	0.7–2.3	0.9–2.5

Derived from Nakos (1984) and Papamichos (1979)
*Average value

Table 4.17 Morphological and analytical data

Horizon	Depth (cm)	Clay (%)	Silt (%)	Sand (%)	Grav. Stones	Structure	Org.mat. (%)	C/N (%)	pH
A	0–10	8	45	47	abundant	wfsbk	7.0	19	5.1
Bw1	10–50	18	35	47	abundant	mmsbk	2.3	12	4.3
Bw2	50–120	21	39	40	abundant	mmsbk	1.6		4.4
C	120+								

Horizon	Color Munsell	Exchangeable cations meq/100 g soil			CEC meq/100 g soil	BS (%)	Extract P (ppm)	Total P (ppm)
		Ca	Mg	K				
A	10YR5/4	12	1.8	0.9	24	61	6.4	450
Bw1	10YR5/4	5	1.2	0.4	21	31	4.4	340
Bw2	10YR6/6	4	1.8	0.4	22	25	3.2	

By: Ministry of Agricultural Development and Foods, Greek Forest Service, data base

Table 4.18 Morphological and analytical data

Horizon	Depth (cm)	Clay (%)	Silt (%)	Sand (%)	Grav. Stones (%)	Structure	Org.mat. (%)	N (%)	C/N (%)	pH
O	3–0						38.7	1.2	19	6.7
A	0–4	21	41	38	0	Sfsbk	14.8	0.7	13	6.3
Bt1	4–39	36	35	29	20–50	Smablk	4.0	02	12	5.5
Bt2	39–65	40	29	31	20–50	Scablk	1.1			5.9
Bt/C	65–86	26	20	54	50–90	Scablk	0.7			5.5

Horizon	Color Munsell	Exchangeable cations meq/100 g			CEC meq/100 g	BS (%)	Extract P (ppm)
		Ca	Mg	K			
O		46.0	4.3	1.5	83.8	62	51.6
A	7.YR3/2	30.5	3.3	1.1	60.0	58	21.5
Bt1	5YR4/3	18.5	2.3	0.5	38.8	55	2.2
Bt2	5YR3/3	20.0	3.8	0.4	34.5	71	2.2
Bt/C	5YR4/4	20.5	3.7	0.3	35.6	70	

By: Papamichos (1979)

Profile: Leptic Regosol (Hapludalf Lithic Udorthents) (P14-Fig. 2.5)
Region: Pieria Region, central Greece.
Parent material: Sandstone Flysch
Physiography: Mountain slope 50%. Aspect: S
Elevation: 1000 m
Vegetation zone: Mountainous
Land use: Oak (Quercus) (Fig. 4.7 and Table 4.19)

4.5 Soils on Basic Igneous Rocks

Basic igneous rocks are not among the most extensive parent materials of soils in this landform. The most common of them are ophiolite, peridotite, and diabase.

These rocks form soils with slightly acid to neutral pH, rich exchangeable Ca and Mg, and argillic horizon with moderately fine-to-fine texture and moderate base saturation. These properties are favorable for the development, in the undisturbed profiles, of illuvial Bt horizons. Thus, the climax diagnostic horizon is the argillic with a reddish color and usually vertic characteristics, due to the presence of expanding clay minerals. In some occasions, mollic surface Ah horizons are present on undisturbed profiles.

Through the sub-Mediterranean to mountainous zones, the climax soil profiles are classified as Vertic Haploxeralfs, Typic Rhodoxeralfs, and Typic/Mollic Udalfs (Vertic Luvisols, Chromic Luvisols, and Rhodic Luvisols).

Soil erosion has affected the greatest part of this landform. However, the fine texture of the soil material has provided them with moderate erodibility. Thus, soil depth in most cases

Fig. 4.7 Typical profiles formed on flysch parent material classified as Typic Xerochrept located in the island of Crete (P15-Fig. 4.3)

is moderate. Parts of the argillic horizon have been preserved in some cases, whereas in others, the profiles consist of materials of altered and/or unaltered C horizon overlying the bedrock. The corresponding soil profiles are classified as Vertic and Typic Ochrepts or Typic/Lithic Orthents (Vertic Cambisols, Eutric Regosols, and Leptosols).

Table 4.19 Morphological and analytical data

Horizon	Depth (cm)	Clay (%)	Silt (%)	Sand (%)	Grav. Stones (%)	Structure	Org.mat. (%)	N (%)	C/N (%)	pH
A	0–12	20	44	36	20–50	Wsbk	5.9	0.17	20	5.3
C	12–34	26	40	34	50–90	massive	1.6	0.06	16	4.8
R	34+									

Horizon	Color Munsell	Exchangeable cations meq/100 g			CEC meq/100 g	BS %	Extract. P (ppm)
		Ca	Mg	K			
A	10YR5/4	5.0	1.8	0.4	24.3	30	3.0
C	7.5YR5/6	1.8	1.3	0.3	20.4	16	0.0

By: Papamichos (1979)

Table 4.20 Ranges of mean values of analytical data of soil profiles on basic igneous rocks

Horizon	Clay (%)	pH	OM (%)	N (%)	C/N	Total P (ppm)	Extr. P (ppm)
O		5.4–6.7	33.5–49	1.0–1.2	18–24	722–742	27.8–48
A	22–28	4.8–6.9	3.6–8.8	0.18–0.36	14–22	285–455	7.4–22.8
B	37–44	5.5–6.6	1.2–3.6	0.07–0.12	10–12	224–357	9.2–17.7
C	22–41	4.6–6.8	0.8–5.5	0.05–0.10	10–13	197–283	3.2–17.0

Horizon	BS (%)*	EDTA Extractable elements in ppm						
		Ca++**	Mg++	K+	Fe	Mn	Zn	Cu
O	53–64	28–42	8.1–13	1.5–1.9	140–662	190–328	18–22	2.4–3.0
A	55–80	10–21.2	8.6–12.4	0.4–0.7	46–225	34–238	1.7–2.0	3.5–3.7
B	28–67	3.0–23.7	2.9–22.2	0.2–0.5	41–155	39–252	0.4–1.8	3.0–5.0
C	34–76	1.5–19.2	1.9–22.6	0.2–1.0	29–187	22–137	0.7–1.7	1.6–2.5

Derived from Papamichos (1979), Nakos (1984), and Economou et al. (1993)

4.5.1 General Soil Properties

The moderate soil depth, the fine texture, the abundance of basic cations, and the favorable pH render the majority of these soils as moderately fertile and productive substrate for forests and pastures. On the contrary, thin soils such Lithic Orthents (Leptic Regosols and Leptosols) developed on serpentinized peridotites become unproductive because of the Mg–K interaction, which causes severe K deficiencies in plants blocking their growth. This phenomenon is extensive in shallow soils on Serpentine in the sub-Mediterranean zone.

Tsatiris et al. (2008) have observed that organic matter, cation exchange capacity, exchangeable bases, base saturation, and the exchangeable Mg/K ratio increase with increasing serpentinization of the parent rock.

Yassoglou (personal observations) has found that the adverse effects on the potassium uptake and growth of pine and olive trees are not evident on soils in which the Bt horizon had not been removed by erosion. These effects were absent also along the contact lines of serpentinized rock with limestone.

Ranges of average values of analytical data and detailed analytical and morphological data of a profile developed on the basic igneous rocks are presented in Tables 4.20 and 4.21.

Profile: Chromic Luvisol (Typic Hapludalfs)
(P16-Fig. 2.5)
Region: Pieria Region, central Greece.
Parent material: Ophiolite
Physiography: Mountain slope 20%. Aspect: NE
Elevation: 1020 m
Vegetation zone: Mountainous
Land use: Oak (Quercus) (Fig. 4.8)

4.6 Soils on Volcanic Rocks

Volcanic rocks are grouped into three main geographical locations of Greece: (a) the volcanic arc of southern Aegean islands, (b) the island of Lesvos in the northern Aegean, and (c) the northern sections of Thrace and Central-Eastern Macedonia in the vicinity to the border line with.

Table 4.21 Morphological and analytical data

Horizon	Depth (cm)	Clay (%)	Silt (%)	Sand (%)	Grav. stones (%)	Structure	Org.mat. (%)	N (%)	C/N (%)	pH
O	2–0						23	0.71	19	6.0
A	0–6	26	41	33	Grav. 20	mfgr	7.5	0.23	19	5.7
AE	6–22	26	39	33	Gr & St. 20	mfgr	2.4	0.09	15	5.9
AB	22–42	29	32	35	Stones 50	mfsbk	1.7	0.08	12	6.0
Bt	42–72	59	24	37	Stones 50	mmsbk				

Horizon	Color Munsell	Exchangeable cations meq/100 g			CEC meq/100 g	BS (%)	Extract P (ppm)
		Ca	Mg	K			
O		1.7	6.3	0.8	44.5	54	4.0
A	10YR3/3	7.5	6.5	0.3	19.5	74	8.4
AE	7.5YR4/2	4.0	17.8	0.1	23.0	52	10.8
AB	7.5YR4/2	2.8	7.6	0.2	19.2	55	22.6
Bt	5YR4/6	3.5	7.7	0.4	34.2	71	

By: Papamichos (1979)

Most of the volcanic formations in Greece belong to the group of acid to intermediate rhyolites, rhyodacites, dacites, andesides, trachyandesides, trychites, and pyrotrchytes and to pyroclastic tuffs and ignimbrites with their ages mostly ranging from the Quarternary to Oligocene. A rather small area is coved by basic volcanic rocks such as basalt and dolerite (Bornovas and Rondoyianni-Tsambaou 1983; Innocenti et al. 1981).

The properties of the majority of the volcanic soils in Greece do not qualify to the criteria of the Andisols as defined by Soil Taxonomy (Soil Survey Staff 1992). In fact, values of their fine earth fractions of (a) ammonium oxalate extracts of aluminum plus iron, (b) the bulk density at 33 kPa retention moisture, and (c) phosphorus retention fall below those which required for andic properties (Georgoulias 2008; Moustakas and Georgoulas 2005).

Genetically, the volcanic soils of Greece show three distinct levels of profile development as indicated by their diagnostic horizons (Yassoglou et al. 1969; Georgoulias 2008; Moustakas and Georgoulias 2005):

a. Soils without horizon differentiation. These soils do not show any genetic process other than the accumulation of small amounts of organic matter on their surface. Thus, their profiles are characterized as Ah, C. The lack of genetic differentiation is due to their extremely coarse texture that does not allow any worth while retention of water in their solum. A secondary reason might be the aridity of the climate. They belong to the southern part of the southern Aegean volcanic arc and are located in the Mediterranean climatic zone and have a xeric soil moisture regime
b. Soils which have developed a cambic Bw diagnostic horizon in a characteristic ABC profile. They are found in the southern Aegean volcanic arc and in the island of Lesvos. Most of the soils have eutric character, but there are also some with dystric. The property which differentiates them from the soils of the above category is the greater age and the higher amount of clay in the parent material (Georgoulias 2008). According to this author's opinion, the greater amount of clay in the parent material is the crucial factor of the genetic differentiation of the profile. This condition allows soil genesis processes to some weathering of the parent materials. This group of soils is located in Mediterranean zone, and its members have a xeric soil moisture regime.
c. Soils having developed an argillic Bt horizon. They have an eluvial A surface horizons and illuvial Bt, overlying the C horizon. They occur at high elevations under udic climatic conditions in the northern sections of East Macedonia and Thrace. There exist cases of dystric conditions which inhibit clay illuviation in the profile (Yassoglou et al. 1969). Under these conditions, a dystric ABC profile is developed, with occasionally an Umbric epipedon.
d. Eroded and occasionally desertified volcanic soils are common in the dry western section of the island of Lesvos and in other Aegean Islands. They usually are characterized by an AC lithic Orthents profile.

4.6.1 General Soil Properties

The area covered by this landform in Greece is relatively small and its importance as a resource is limited. Their basic morphological, physical, and chemical properties are given in Table 4.21.

Fig. 4.8 A typic Rodoxeralf (*up*) and a Lithic Xerorthents (*down*) on basic igneous rock in the island of Aegean Sea

Profile: Haplic Luvisol (Typic Hapludalfs) (P17-Fig. 4.3)
Region: East Macedonia, northern Greece.
Parent material: Dacite
Physiography: Mountain slope 20%. Aspect SW
Elevation 1200
Vegetation zone: Mountainous
Land use: Beech forests

Even though the criteria for the definition of andic properties were not clearly defined at the time the above data were obtained, the ammonium oxalate extractable Al is by far smaller than the presently accepted limit. Therefore, it is concluded the data of the soil profile of the Table 4.21 do not qualify as andic.

Another soil profile on dacite under grass vegetation, in the same landform, examined in the same investigation (Yassoglou et al. 1969) had developed a more acid profile with an Umbric surface epipedon, overlying cambic horizons with low base saturation and no clay illuviation. Thus, it was classified as Typic Haplumbrept.

Table 4.22 Morphological and analytical data of a Haplic Luvisol (Typic Hapludalf) (Yassoglou et al. 1969)

Horizon	Depth (cm)	Clay (%)	Silt (%)	Sand (%)	Grav. Stones (%)	Structure	Org C (%)	C/N (%)	pH
A1	0–3	20.2	20.6	59.2	0.0	Smsbk	6.16	19.9	5.1
A2	3–10	20.5	17.3	62.2	0.0	Smsbk	2.46	20.5	5.0
A3	10–12	19.2	19.2	61.6	2.4	Mmsbk	1.11	13.7	5.3
Bt1	18–70	31.8	13.3	54.9	9.3	Msbk	0.61	7.6	5.4
Bt2	70–95	17.3	13.1	69.6	8.8	Wmsbk	0.16	4.0	5.4
C1	95–145	11.0	12.1	76.9	28.6	Sgr	0.09	4.5	5.3
C2	145+	8.4	13.3	78.3	56.1	Sgr	1.06	3.2	5.8

Horizon	Color Munsell	Exchangeable cations meq/100 g		CEC meq/100 g	BS (%)	Extractable	
		Ca	Ca/Mg			Fe_2O_3 (%)	Al_2O_3 mg/100 g
A1	10YR2/2	13.7	4.8	36.8	48	0.54	2.8
A2	10YR3/2	5.4	3.6	23.1	33	0.71	33.1
A3	10YR4/3	1.9	2.2	18.5	17	0.81	86.8
Bt1	10YR3/3	1.9	1.6	15.4	23	1.16	73.7
Bt2	10YR5/6	4.6	2.4	11.0	62	0.78	44.6
C1	7.5YR5/8	4.6	2.7	10.6	62	0.76	33.1
C2	10YR7/2	4.0	2.0	9.2	65	0.36	34.4

In a further investigation of the above two soil profiles, Apostolakis and Douka (1970) found significant differences in the plant nutrient contents of the profiles, probably affected by the difference in the vegetative cover and lower pH. Total N, exchangeable K, and extractable Cu were greater under grasses than under beech, whereas extracted Zn was higher under beech and extractable P about equal under both vegetative covers.

A morphological description and analytical data of a typical soil profile in the southern Aegean vocanic arc are given in Table 4.22. The soils in the islands of the Aegean Sea have been extensively investigated by Moustakas and Georgoulias (2005) and Georgoulias (2008). They have concluded that they do not contain enough non-crystalline amorphous materials, and they do not have low enough bulk density and phosphorus retention capacity to be characterized as soils with andic properties. Coarse-textured soil profiles on the youngest volcanic materials have not been genetically differentiated due to their young age and the coarseness of the parent material. Soil profiles with clay contents greater than 18% seem to undergo moderate profile differentiation developing Bw cambic subsurface horizons. This, however, needs further verification. The minerals that dominate the clay fraction are of the 2:1 type of the smectite group.

Cultivated volcanic soils are mainly those of the Aegean Islands. They are of relatively low productivity and highly vulnerable to erosion and desertification in the drier regions. A large portion, in the dry west region of Lesvos, has been severely eroded. They have shallow depths due to overexploitation. The soils at the present state are classified as Lithic Xerorthents or Leptosols with very low productivity. In spite of their low productivity, some of them in the southern Aegean volcanic arc produce small but of high-quality products such as wine, fruits, and vegetables (Fig. 4.9).

Profile: Arenic Regosol (Vitrandic Xeropsaments) (P20-Fig. 4.3)
Region: Thera Island, Aegean Sea, Greece
Parent material: Upper rhyodacitic pumice
Physiography: Pleistocene terrace, slope: 0–2%. Aspect NE-e
Elevation: 150
Vegetation zone: Mountainous
Land use: Wild herbs and shrubs (Table 4.23)

4.7 Tertiary Deposits

Tertiary deposits of Greece are either siliceous or calcareous depending on the original hard parent rock. They are usually moderately deep to deep and are located within the Mediterranean and sub-Mediterranean vegetation zones, at elevations usually ranging between 500 and 1000 m (Nakos 1984).

Both are highly erodible and have severely truncated profiles on slopy terrains, which have lost their protective

Fig. 4.9 A moderate deep Typic Xerochrept (*up*) soil profile and a tube soil (*down*) located on lava parent material in the island of Lesvos (P18-Fig. 4.3)

Table 4.23 Morphological and analytical data of a Vitrandic Xeropsaments (Moustakas and Georgoulias 2005)

Horizon	Depth cm	Clay (%)	Silt (%)	Sand (%)	$CaCO_3$ (%)	Structure	Org C (%)	C/N (%)	pH
A	0–15	3.9	19.3	76.8	0	Structureless	2.14		8.60
C1	15–32	2.7	17.7	79.5	0	≫	0.69		8.89
C2	32–57	2.7	19.3	78.1	0	≫	0.64		8.78
C3	57–79	3.8	26.1	70.1	0	≫	0.48		8.80
C4	79+	4.4	25.9	69.7	0.77	≫	0.52		8.97
Horizon	Color Munsell	Exchangeable cmol(+)Kg^{-1} soil				CEC meq/100 g	ESP (%)	Extractable in wt. %	
		Ca	Mg	K	Na			Alox	Feox
Ah	10YR6/2	5.25	1.67	0.563	0.314	8.03	3.91	0.034	0.026
C1	10YR6/3	3.50	1.60	0.358	0.492	7.09	6.94	0.29	0.018
C2	10YR6/3	2.50	0.80	0.255	0.807	6.50	12.42	0.027	0.012
C3	10YR6/3	3.25	1.83	0.339	1.240	8.10	15.31	0.032	0.011
C4	10YR6/3	2.25	1.97	0.397	1.270	7.56	16.80	0.034	0.012

Table 4.24 Ranges of mean values of clay content and chemical properties of soil horizonsin profiles of mountainous and hilly soils on silicious tertiary deposits

Horizon	Clay (%)	pH	OM (%)	N (%)	C/N	CEC* meq/100 g	Total P (ppm)	Extr. P (ppm)
O		6.2–6.6	32.5–36.2	0.92–1.6	18–21		750–867	270–50.4
A	22–24	5.7–6.1	4.1–4.9	0.16–0.18	16–18		310–356	7.2–10.6
B	30–40	5.3–5.9	1.0–1.55	0.07–0.05	11–15		215–362	3.4–17.0
C	29–45	5.3–6.1	0.49–0.84	0.04–0.05	8–11		331–347	9.6–10.6
Horizon	BS (%)*	Exchangeable cations meq/100 g			EDTA Extractable elements in ppm			
		Ca++**	Mg++	K+	Fe	Mn	Zn	Cu
O	59–62	34.0–43	5.4–6.3	2–02.1	36–168	134–833	27–30	2.3–3.1
A	40–54	15–12	3.5–4.0	0.6–0.64	34–177	73–223	26–31	2.1–2.2
B	53–35	6.7–18.4	1.8–4.9	0.42–0.52	19–82	100–62	0.8–1.0	2.4–2.6
C	40–57	7.3–19.5	4.3–5.2	0.32–0.42	14–76	61–37	0.1	1.6–2.2

Derived from Nakos 1984

vegetative cover. Siliceous tertiary deposits differ in their properties as they are affected by the presence or absence of carbonates (Table 4.24).

4.8 Soils on Man-made Terraces

Soils on stony terracing constitute a specific category of soils build on both residual and allochthonous soils. Their construction dates since several centuries ago on hilly areas of Greece, in order to allow degraded sloping areas to be effectively protected from erosion and profitably cultivated (Fig. 1.1). Soil material has been transported from the surrounding areas, mixed, and laid into the terraces. The dominant soil-forming factor has been the extreme anthropic interference. New young soil profiles were formed in the new man's affected parent material. These profiles are completely different from the adjacent usually shallower soils. The properties of these soils heavily depend on age of the terrace, the type of vegetation, the land management practices, and the prevailing climatic conditions. Land management practices such as planting particular species, adding manure, diverting rain runoff water into the terrace, etc., were applied on these particular categories of soils. Today, these constructions support in certain cases of valuable soil for growing perennial or annual crops, providing important income for the farmers on less favorable areas of Greece, supporting high biodiversity of flora and fauna, and protecting sensitive areas from degradation (Fig. 4.10).

Soil properties of a well-managed forest nursery on terracedsloping land are reported in Table 4.25.

Agricultural terracing by redistributing soil materials creates flat or relatively flat strips of land to allow cultivation

Fig. 4.10 Old style terracing land with linear (*up*) or individual (*down*) terraces build with stones (*Photograph* by C. Kosmas)

Table 4.25 Surface soil properties of a forest nursery constructed on a flyschsloping land (Yassoglou et al. 1969)

Texture	pH	Organic matter (%)	N (%)	C/N	CEC meq/100 g soil	B/S (%)
Loamy	5.2	4.9	0.27	10.5	20.0	50
Extr. Cu ppm	Extr. K ppm	Extr. Mn ppm	Extr. Fe ppm	Extr. P ppm	Extr. Zn ppm	Carbonates %
1.5	290	77	1.5	52	3.5	0.0

of sloping areas. Land terracing is a very old technique in Greece and else where for cultivating hilly areas.

terracing ofsloping land (a) increases plant root depth, (b) makes a sloping land less steep, improving access to farming operations, (c) reduces stoniness, and (d) increases the absorption of water by the soil in heavy rainfalls reducing soil erosion. There are various techniques in land terracing depending on the existing technology. In the present discussion, only the old style of using stones for building terraces has been considered. Pocket, braided, or broad-base terraces were constructed in areas with shallow soil overlying a consolidated bedrock. In such cases, stones were transported by the animals for building the walls in selected sites in the hilly slope, and then, the terrace was filled with soil from the area surrounding the site.

The art of terracing is nearly as old as farming itself considered. Archaeological evidences indicate that in Argolida, soil erosion was controlled by terracing in the late Bronze Age (Grove et al. 2001). There are ancient terraces in the majority of Aegean islands and in Crete island. Hilly areas under olive groves in the island of Lesvos were terraced using stones several hundred years ago, especially during the Byzantine and Ottoman period. The olive groves in this island cover an area of 60,966 ha. This area can been considered as a man-made landscape supporting valuable soil and allowing the growth of olive plantations with a biodiversity even higher than adjusted pine or oak forests.

Erosion within a terrace unit is a complex series of processes. Soil spots with high erosion rates are usually accompanied and balanced with spots of deposition. When the soil surface in the sloping part of the terrace is not well protected by vegetation, water runoff can be generated when

Fig. 4.11 Stony terraces collapsed due to removal a stone in the base (*up*) or trampling of animals through the land (*down*)

a heavy rainfall occurs transporting soil from the sloping part and depositing to the almost flat part of the terrace.

In recent decades, the value of terraces has markedly declined due to: (a) difficulties associated with accessibility and use of machinery, (b) decreasing price of agricultural products and increasing labor costs, (c) high-input agriculture developed in plain areas, and (d) extensive migration of people from rural to urban areas.

The non-irrigated cultivation of the terraces was adapted to the power of men and animals, or to the small machines, mechanical donkeys that served farmers until the mid-twentieth century. Once terracing land is abandoned, as unprofitable, the collapse of stony terraces may be rapid and soil can be immediately washed out by surface water runoff and gravity, leading to accelerated land degradation (Fig. 1.2). Usually, maintenance of terraces is nothing more than replacing fallen stones (Fig. 4.11).

Detailed studies on terrace collapse in the island of Lesvos have shown that the stability of terraces is mainly related to the slope gradient, soil type, stone composition, and management practice (Kosmas et al. 1998). Another important factor affecting terrace collapse is the degree of shrinking–swelling measured by the coefficient of linear extensibility (COLE) of the soil into the terrace. Soils with high amounts of expanding clays such as vermiculite or montmorillonite exert horizontal pressure after soil wetting is high, especially at the base of the terrace, causing inflation, instability, and finally terrace collapse.

Terracing on unconsolidated or slightly consolidated sloping tertiary landscapes in Greece proved to be difficult to be protected from severe erosion and gullying.

Slope gradient also greatly affects terrace structure stability. Detailed studies carried out in the terracing land of Lesvos island showed that there are three critical slope gradient classes: (a) less than 15%, (b) 15–35%, and (c) greater than 35% (Fig. 4.1). Stony terraces constructed on slopes less than 15% can be characterized without risk remaining almost undisturbed under the prevailing management conditions in the area (Kosmas et al. 1998). The rate of collapse increases almost linearly with increasing slope gradient from 15 to 35%. This slope class can be characterized as fragile in relation to the rate of collapse. The rate of collapse is very high on slopes greater than 35%. This area can be characterized as critical, and special attention is required on land management (Fig. 4.12).

The climax diagnostic horizon of the soils formed on terraces is the cambic horizon. In some cases, a mollic epipedon has been formed especially in case that the soils are under perennial crops such as olives. The dominant climax

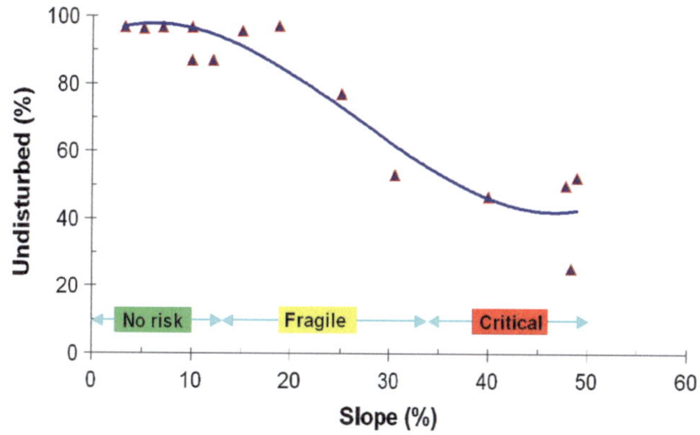

Fig. 4.12 Relation of terrace stability with slope gradient in the olive groves of the island of Lesvos (Kosmas et al. 1998)

taxonomic unit is Xerochrept (calcaric, eutric Cambisol) or Xerorthents (calcaric, eutric Regosols).

As a conclusion, it could be stated that terracing of sloping land results in a naturally unstable landscape, of which the preservation requires consumption of significant energy, man power, and financial resources. At the same time, the cost/benefit ratio under the present socioeconomic conditions of Greece does not, in most cases, favor expansion of land terracing. Repairing and maintenance of existing old terracing should be carefully evaluated in terms of expected socioeconomic cost and benefits. The cost of inaction, in terms of land deterioration, should also be taken into account.

It seems that preservation of existing terracing of relatively deep soils on moderately sloping land under permanent productive plant cover could be recommended. Such are cases of olive groves and vineyards on moderate slopes.

In the following chapter are described soils of Greece, of which the parent materials have been transported from other landscapes mainly through the gravitational flow of surface waters. Less important modes of transportation have been Aeolian (wind) transportation and landslide movements.

Many of the transported materials have been originated from developed soils of the upper lands, which at certain stage of their evolution suffered various degrees of truncation caused by erosion and translocation to lower lands. Thus, transported soils located near the origin have retain, at various degrees, the properties of the original soils. However, fluvial soils, deposited by big streams and rivers originated from large watersheds, have highly highly mixed parent materials.

The following are the categories of soils of Greece with transported parent materials.

5 Colluvial Soils

Colluvial soils consist of locally transported detritus materials of soil horizons and parent materials of sloping terrains from the upper sections of the slopes through water erosion or landslides. They have been deposited at the foot of the slope or at various parts of it, where the local relief impedes the further surface flow of the detritus material.

These soils are among the most productive of the mountainous and hilly landscapes because they have significantly deep rootable material, which provides moisture and nutrients to the plants. Their disadvantage is their vulnerability to erosion and landslides. The latter are more frequent on slopes with Flysch and Tertiary clay. These soils are found as inclusions covering small areas mostly within the soils in clay deposits and conglomerates (next Chap. 6), and thus are not discussed in detail.

N. Yassoglou et al., *The Soils of Greece*, World Soils Book Series,
DOI 10.1007/978-3-319-53334-6_5

6 Soils on Clay Deposits and Conglomerates

Clay deposits of Quaternary and Tertiary age were erodible residual soil and geological materials, which have been transported from original multiple local watersheds and deposited on lower lands. They are associated with depositional environments such as continental formations, large lakes, or marine basins. Conglomerates consist of various mixtures of Quaternary fine earth materials (materials with a diameter less than 2 mm) and well-rounded rock fragments. The upper part of these materials is usually unconsolidated and has formed soils of various thicknesses. The lower part of these deposits has often been consolidated with various cemented materials such as silica and carbonates. Similar to clay deposits, conglomerates have been deposited in the periphery of water bodies (lakes or marine basins). Deposition of these materials took place at short distances, without being sorted according to their depositional order. Thus, they consist of mixtures of materials with various contents of rounded rock fragments and fine earth. Soils on these parent materials cover a significant part of Greece estimated in 2,067,000 ha or 15.6% of the total land (Fig. 6.1). These soils are located in various parts of the country and especially in northern and western Peloponnesus, Attica, eastern Thessaly, central and western Macedonia, and the islands of Evia, Crete, and Rodos.

Clay deposits or conglomerates have often been deposited over marl deposits. However, the consequential removal of these materials by soil erosion has exposed the marls on the soil surface (Fig. 6.2). The majority of these soils are on hilly areas, sensitive to degradation due to soil erosion, especially in case of overcultivation and low plant cover. The soil color is usually red or yellow-red inherited mainly from the parent material from which these materials have been formed. The majority of coarse fragments exiting on these deposits have limestone origin, which favors the formation of the iron oxide hematite under well-drained conditions (Yassoglou et al. 1997).

The rate of erosion is largely affected by the amount and the size of rock fragments on the soil surface. Experimental studies in Greece have shown that rock fragments in the dimension of pebbles (diameter 2–75 mm) on the soil surface can reduce significantly surface water runoff and soil loss playing an important role on land protection from degradation (Moustakas et al. 1995). In addition, rock fragments in the dimension of cobbles (diameter 75–250 mm) restrict evaporative water loss during periods of moderate drought (e.g., from late fall to early summer), but they increase evaporation during the dry and hot summer (Danalatos et al. 1995). Furthermore, soils on conglomerate parent material containing considerable amount of rock fragments become warmer earlier in spring than the same soils free of rock fragments. This favors early plant growth and better use of available water.

The landscape of these deposits is usually highly dissected, due to high rates of rainwater runoff. Various channels are crossing these areas forming a mixed landscape of plateau with steep sloping areas along the channels. The high density of drainage network is attributed to the low infiltration rates of soils due to both usually high amount of clay content reducing water infiltration rate and low soil porosity of sublayers that are highly compacted reducing hydraulic conductivity to downward percolation of rainwater. The prevailing adverse topographic characteristics in combination with low infiltration rates of water have favored high soil erosion rates in the past. As shown in Fig. 6.3, the subsurface calcic horizon, which is usually formed in a depth greater than 75 cm, under the Greek climatic conditions, has been exposed throughout the soil surface.

Soils formed on clay deposits or conglomerate parent material contain high amounts of 1:1 clay minerals such as kaolinite. However, some 2:1 clay minerals such as vermiculite and montmorillonite are present in low quantities. This is mainly inferred from both the absence of cracks on the soil

N. Yassoglou et al., *The Soils of Greece*, World Soils Book Series,
DOI 10.1007/978-3-319-53334-6_6

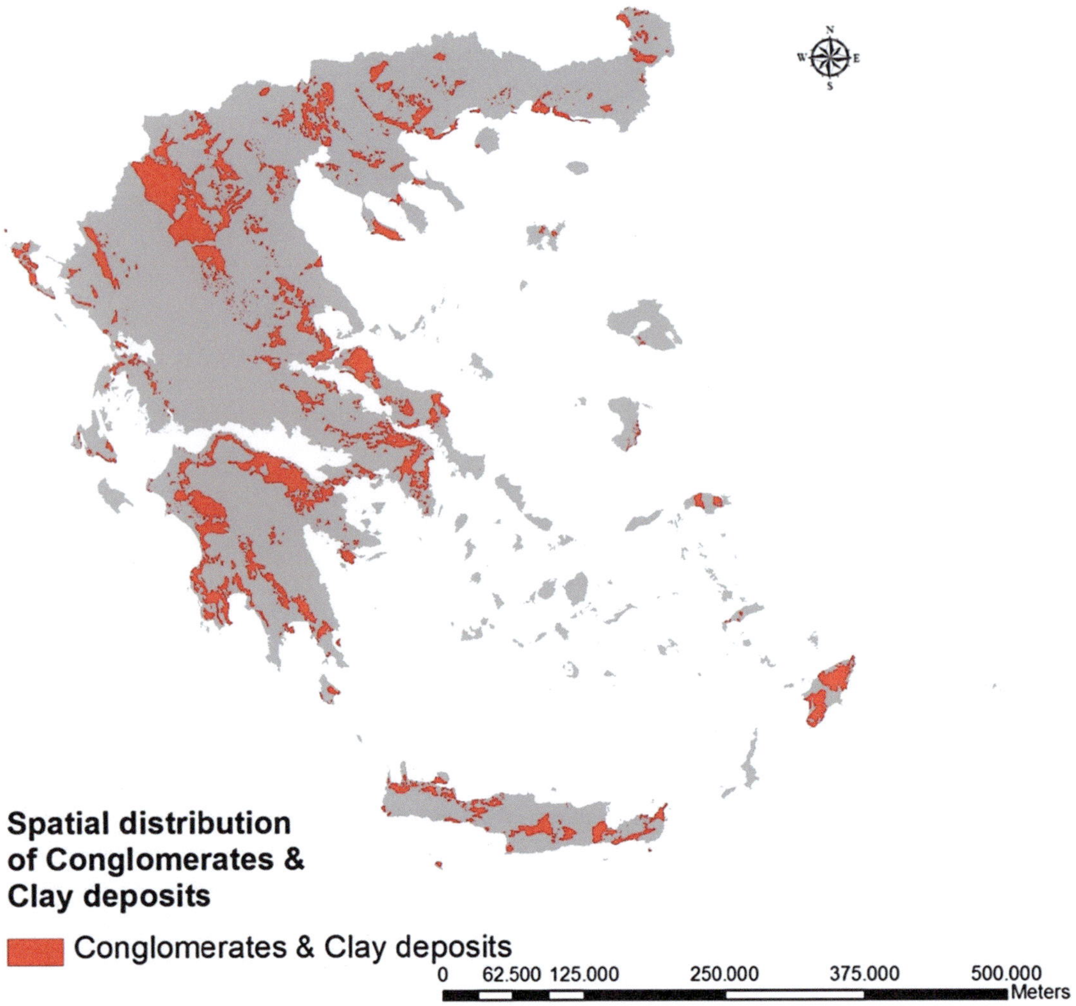

Fig. 6.1 Spatial distribution of soils formed on clay deposits and conglomerate parent materials

surface during the dry period and mineralogical analysis of selected soils. Therefore, the soils formed on these parent materials are characterized by moderate productivity.

The soils are mainly classified as Inceptisols (Cambisols) in 79.9% of this land, while the next important taxonomic unit is Alfisols (Luvisols) in 13.2%. Other taxonomic units such as Regosols and Leptosols have been identified in areas covering 2.8 and 3.2% of this land, respectively. The climax diagnostic horizon of the soils formed on clay deposits and conglomerates is cambic and argillic horizons. Cambic horizon is present in less stable areas, while in areas with limited soil erosion, an argillic horizon predominates. The dominant climax taxonomic units are Xerochrept (calcaric Cambisol, calcaric Regosols) or Rodoxeralfs and Haploxeralfs (eutric or rhodic Luvisol). Regosols (Xerorthents) are formed in highly eroded areas with large amounts of rock fragments.

6.1 General Soil Properties

Semi-detailed soil survey studies (Kosmas et al. 2010), carried out in the majority of the soils formed on clay deposits and conglomerate parent materials, have shown that the dominant textural classes are moderately fine (clay loam, sandy clay loam) or fine (clay, silty clay) at 54.5 and 36.9% on the clay and conglomerate landscapes, respectively. However, some medium- (loam) or coarse-textured soils have been identified in small percentages (4.0 and 2.8%, respectively).

The soils are mainly characterized as moderately deep (soil depth 60–100 cm), deep (soil depth 100–150 cm), and very deep (soil depth > 150 cm) in 35.9, 22.9, and 23.7% of this land, respectively. Shallow (30–60 cm) to very shallow soils (soil depth < 30 cm) have been found in 14.1 and 2.9% of the land, respectively. Sometimes, these soils have

Fig. 6.2 Typical landscape of clay deposits and conglomerate materials subjected to soil erosion due to intensive cultivation

limiting layers such as petrocalcic horizon, behaving like hard rock or gravels in 15.8 and 13.7% of this land, respectively.

Concerning drainage conditions, the soils are mainly very well to well-drained covering 59.6 and 26.8% of this land, respectively. Some soils are somewhat well-drained or poorly drained in 11.3 and 2.3% of the area, respectively.

The soils either are free of rock fragments or contain high amounts of rock fragments. Areas with rock fragment less than 20% cover 53.6% of landscape, while the classes of rock fragments "high" (20–60% RF) and "very high" (>60% RF) cover 30.9 and 15.5% of the total land, respectively. The high amounts of rock fragments in some cases are attributed to high amount in the parent material or to high erosion rates favoring the concentration of rock fragments in the soil surface.

The dominant slope gradient classes are the slightly slopping (slope 2–6%) and moderately sloping (slope

Fig. 6.3 Severely eroded soil formed on clay deposits with the calcic horizon almost exposed in the soil surface

6–12%) covering 54.5 and 24.2% of this landscape, respectively. However, nearly flat areas (slope 0–2%) or steeper slope gradients than the previous classes can be identified such as strongly sloping (slope 12–18%) in 15.4 and 5.1% of this land, respectively.

Concerning the degree of soil erosion, they are usually characterized as slightly or moderately eroded, covering 42.1 and 35.9% of the landscape, respectively. However, severe or very severe eroded soils can be found in steep areas with low percentage of plant cover covering less than 13.5% of this land. No eroded soils have been described in 9.1% of the area of these lands.

Inorganic carbonates, detected in the field by using hydrochloric acid, are usually present in the whole soil profile covering 61.2% of the total area of land. Soils in which pedogenetic processes were active for long period have lost inorganic carbonates in the upper soil layers or in the whole profile in the 28.8% of the area of this landform.

Ranges of average values of soil properties of the soils formed on clay deposits and conglomerates parent materials over the various bioclimatic zones of Greece and data of typical profile are given in Tables 6.1, 6.2, and 6.3.

Profile: Eutric Luvisol (Typic Haploxeralfs) (P21-Fig. 2.5)
Region: Northern Peloponnesus, Greece
Parent material: Stratified clay deposits
Physiography: nearly flat (slope 2%) Aspect: plain
Elevation: 58 m
Vegetation zone: Submediterranean
Land use: Olives

Profile: Calcaric Cambisol (Typic Xerochrepts) (P22-Fig. 2.5)
Region: Epirus, northwestern Greece
Parent material: Stratified conglomerates
Physiography: Moderately slopping (slope 12%) Aspect: SE
Elevation: 28 m
Vegetation zone: Submediterranean
Land use: Olives

This landform belongs to the land capability classes I, II, and IIIdue to low, moderate and strong erosion risk limitations (Kliengebiel and Montgomery 1961). Classes I and II are the dominant. The respective land suitability classes are the S_2 and S_3 (FAO, 1976) and S_3, because of their soil erosion, shallowness and drought limitations. Biomass productivity and tolerance to external damaging influxes are moderate. The fertility of these soils is moderate with nitrogen being the element in insufficient quantities.

Table 6.1 Ranges of mean values of clay content and chemical properties of soil horizons in profiles of soils formed in marls mainly in cropland areas

Horizon	Clay %	pH	OM %	N %	C/N	Extr.P (mg/kg)
Ap	6.7–74.7	4.4–8.5	0.4–6.3	0.17–3.00	8.6–11.4	1.1–304
B	4.5–77.5	4.7–8.9	0.2–4.3	0.15–2.40	9.9–13.8	0.4–246.6
C	8.7–67.5	4.7–8.7	0.1–2.3	0.05–2.00	8.9–13.3	0.4–172

Derived from: OPEKEPE 2014

Table 6.2 Morphological and analytical data

Horizon	Depth cm	Clay %	Silt %	Sand %	Gravel/Stones (%)	Structure	Org. mat. (%)
1Ap	0–15	45.8	21.0	33.2	5	wmsbk	2.9
1ABt1	15–39	48.2	18.4	33.4	10	wcsbk	0.8
1Bt2	39–64	54.8	21.2	24.0	0	mcsbk	0.85
1Bt3	64–105	51.8	20.0	28.2	0	mcabk	0.6
2C	105–167	51.8	14.0	34.2	5	scabk	0.4

Horizon	Color Munsell	Exchangeable mg/kg			CEC cmol/kg	BS %	Extractable P mg/kg
		Ca	Mg	K			
1Ap	10YR 4/4	1920	120	215	18.1	75	12.3
1ABt1	10YR 4/6	1162	129	89	13.2	77	10.2
1Bt2	10YR 5/6	1167	201	77	16.2	78	7.8
1Bt3	10YR 5/6	1250	248	96	22.1	82	4.3
2C	10YR 4.5/6	1493	233	92	16.1	87	3.3

By: OPEKEPE (2014)

Table 6.3 Morphological and analytical data

Horizon	Depth cm	Clay %	Silt %	Sand %	Gravel/Stones (%)	Structure	Org. mat. (%)	C/N %	Ph
1Ah	0–18	27.7	25.4	46.9	60	sfsbk	2.2	10	7.8
2Bw1	18–39	39.4	23.2	37.4	60	sfsbk	1.1	12.1	7.6
3Bw2	39–70	28.5	30.2	41.3	75	mmsbk	1.4	13.2	7.9
3C1	70–151	25.6	27.8	46.6	80	none	0.5	12.5	7.8
3C2	151–170	25.8	25.9	48.3	65	wfsbk	0.3	11.5	7.9

Horizon	Color Munsell	Exchangeable mg/Kg			CEC cmol/Kg	BS %	Extractable P mg/kg
		Ca	Mg	K			
1Ah	5 YR 4/3	7140	82.5	261	24.5	100	9.1
2Bw1	5 YR 5/6	6299	60.1	205	23.3	100	6.3
3Bw2	5 YR 6/6	6648	64.2	150	18.3	100	8.0
3C1	5 YR 6/6	6402	57.1	189	13.2	100	11.0
3C2	5 YR 6/5	6400	53.3	103	13.2	100	6.2

By: Kosmas and Antonoglou (2002)

6.1.1 Management Groups

The soils formed on clay deposits and conglomerates usually present moderate-to-severe limitations for agricultural development. The main limitations are soil depth, presence of rock fragments, and slope gradient. They are mainly characterized as moderately productive soils. They are mainly cultivated with cereals, cotton, vegetables, olive trees, vines, almonds, and other perennial or annual crops. These soils are suitable for annual crops and moderately suitable for perennial crops. They are very suitable for vineyards. In case that are under natural vegetation, the main plant species are pine forests or oak plantations or Mediterranean shrubby vegetation.

Recent Lacustrine and Organic Soils

7

The lacustrine mineral and organic soils discussed in this section belong to lands, formed in recent lake beds and river deltas. These landforms frequently coexist with recent alluvial deposits. They belong to the lowest toposequence and youngest chronosequence of allochthonous soils of the country. Geomorphologically, these are parts of recent coastal or inland flood plains, at elevations not exceeding the 20 m from them. The origins of the inorganic parent materials are erosion products transported from local, multiple, or larger (such as river) watersheds. Organic soils (Histosols) form in situ under waterlogged conditions. The source of the water is either surface flow or phreatic.

Lacustrine soils on older Quaternary or Tertiary surface occupy significant areas in Greece, especially in Thessaly.

Soils in recent lacustrine depositions contain high amounts of organic carbon, or occasionally they are interlayered with organic deposits. Organic soils have been usually formed in the marshes or lakes, especially in their periphery. Within this zone, water reaches a maximum level of around 50 cm during the wet period. Lacustrine and organic soils are mainly located in Philippi marshes (Macedonia), Xiniada and Kopais (Sterea Hellas), Agoulinitsa (Peloponnesus), and Paramithia and Pamvotida (Epirus). It is estimated that these soils cover an area of about 39,600 ha.

Lacustrine mineral soils contain high amounts of carbonates (usually higher than 40%) due to the following: (a) the presence of calcareous parent materials in the surrounding areas (limestone, marls) and (b) the growth of large amount of shells favored by the wet environment which are deposited after their death.

Lacustrine deposits usually have a plate structure since the sediments have been settled in the bottom of the lake periodically after each flooding event. Plate structure partially restricts the downward flow of water and plant root penetration. Farmers use to plow the soil very deep to break this type of structure.

Lacustrine mineral deposits are usually enriched with organic carbon ranging from 3.4 to 14.8%. The distribution of organic carbon with depth is irregular depending on the depth of the water that each soil layer has been formed. In periods that water depth in the lake was less than 50 cm, hydromorphic plants are growing during the spring period when water table is lowering, and then during the winter period, plants are deposited in the bottom of the lake. In periods that the level of water in the lake increases above 50 cm, inorganic sediments are deposited over the previous organic layer. Their structure and composition is often stratified.

Some of the soils formed on the lacustrine deposits, located nearby the coastal line and inland-drained lakes in the semiarid zone of the country, contain high amounts of soluble salts due to the intrusion of brackish water or repeated evaporation of inland flow. This become apparent during the dry period in which the upper soil layers become dry due to high water evaporation rates. In addition, in case that the soil is not cultivated plate clods are formed in the uppermost soil layer during the dry period due to soil desiccation caused by the high rates of soil water evaporation (Figs. 1.1 and 7.1).

Organic soils usually consist of well-decomposed organic materials characterized as "sapric." In some cases, they have subsurface semi-decomposed fibric layers characterized as "hemic" decomposition of the surface organic matter in the semiarid climatic zones is enhanced by high microbial activity. Preservation of the subsurface organic deposits is favored by the aquic soil moisture regime created under waterlogged conditions, even in the semiarid zone. The thickness of the organic layers is largely variable ranging from less than 1 m to more than 200 m depending on the period of an existing lake. A study carried out by the Greek Public Electricity Company (DEH) in the area of Philippi marshes (Macedonia) has shown that organic materials have been extended from the soil surface to more than 200 m

N. Yassoglou et al., *The Soils of Greece*, World Soils Book Series,
DOI 10.1007/978-3-319-53334-6_7

Fig. 7.1 Lacustrine deposit with plate clods falling apart due to soil desiccation caused by high water evaporation rates (*Island of Lesvos*)

depth. In many cases, organic deposits have been buried by alluvial deposits, when the whole lake has disappeared. An example is given in Fig. 1.2 in which organic deposits have been buried by huge amounts of sediments transported by the river and a wetland has been converted to a pine-forested area.

The majority of the organic soils have been introduced to cultivation by lowering water table after constructing a drainage network (Fig. 7.2). After lowering of water table, oxidation of organic materials is accelerated by the increasing aero-porosity and oxygen content into the soil. Measurements in organic soils in the area of Tenagi Filippon have shown that oxygen content is related to the level of ground water table (Kosmas et al. 2005) (Fig. 7.3). During the winter period when the water table was between 50 and 120 cm, oxygen content was lower than 4%. By lowering the ground water table during the dry period due to subsurface irrigation of the growing plants, oxygen content in soil depth 40–60 cm increases linearly up to 16%. An outlay of points included in Fig. 7.3 show high concentration of

Fig. 7.2 Alluvial deposit-overlaid organic deposit formed in a previous lake under a pine forest (*Island of Lesvos*)

Fig. 7.3 Typical landscape of organic soils (*up*) with a representative profile in the same area (Macedonia) (*down*) (P23- Fig. 4.2)

oxygen, while the ground water table was relatively shallow. These measurements were carried out after a heavy rainfall which enriched the soil with oxygen.

The amount of oxygen content ($W\%$) in relation to the ground water table (Z, m) can be assessed by the following equation:

$$W = a/(1 + e^{(b - c*z)})$$

where a is the maximum oxygen content (%) equal to 19.7, b is the initial value of oxygen content (%) equal to 6.2, c is a constant of the rate of increase of oxygen content (%)

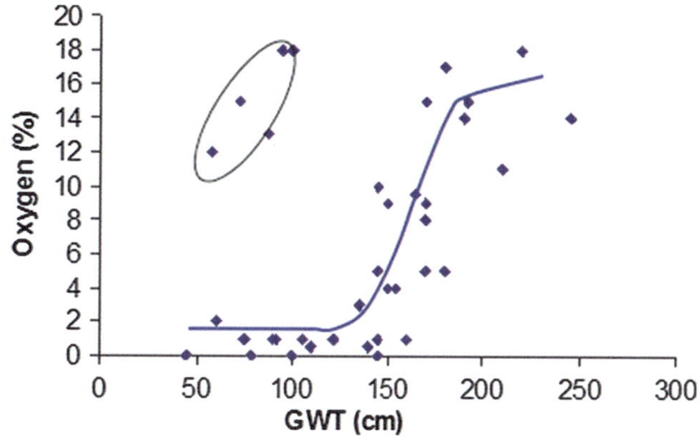

Fig. 7.4 Relation of oxygen content with depth of ground water table in the organic soils of Tenagi Filippon (Macedonia)

equal to 0.039 for the specific area and soil depth of 20–40 cm estimated by the approach Machura and Mulawa (1973). Based on the previous study, it is proposed to keep the water table in the level of 110–120 cm by regulating it in a drainage network, avoiding further oxidation of organic materials, and providing adequate oxygen in the plant roots for normal growth (Fig. 7.4).

The oxidation of organic materials in combination with the use of heavy machineries for cultivation has resulted in a considerable subsidence of organic soils in Greece. Measurements carried out in the organic soils of Philippi have shown a subsidence rate of 3–5 cm per year (Kosmas et al. 2005). Subsidence resulting in considerable lowering of the soil surface bringing the water table on the soil surface (Fig. 6.1) since the drainage network is unable to remove of excess of water by gravity.

Organic soil distribution in Greece is presented in Fig. 7.5.

7.1 General Soil Properties

Organic soils and soils in lacustrine deposits have usually bulk density lower than 1 g/cm^3 depending on the type of organic materials and the stage of decomposition. When oxidation of organic carbon has preceded in advanced stage, bulk density increases and soil porosity decreases imposing limitations in water movement and root penetration. Such conditions have been observed in the upper 35–40 cm of soil layer in organic soils.

Organic and lacustrine soils are saturated with water from late October to early May since they are very poorly drained. Ground water table approaches the soil surface during the wet period, while during the summer period is lowering

Fig. 7.5 Distribution of organic soils in Greece

below to 1.5 m since an upward capillary movement of water is generated by the high evaporation and plant transpiration rates caused by the dry climatic conditions.

The soils are usually nearly flat (slope 0–2%) to gently sloping (slope 2–6%), since they have been formed in the bottom of a lake or in the surroundings of the lake. Therefore, erosion is absent. However, some new materials may be added after heavy rainfall events generating high amounts of runoff in the surrounding hilly areas.

A case study carried out by NAGREF (National Agricultural Foundation) in the frame of the National Soil Survey Greece showed that organic matter content of soils of Greece has declined due to the intensive cultivation. Soils of Greece were grouped in four classes depending on the soil organic carbon content (SOC) (<1.0, 1.0–2.0, 2.0–5.0, >5.0%). The main value of SOC for the topsoils (0–30 cm) of surface area ranged between 26.8 and 221.1 t ha^{-1} (Jandi and Rodegniero 2011).

Lacustrine and organic soils are usually very deep (depth >150 cm) or deep (depth 100–150 cm). In some cases, a limitation layer of sandy materials is present. Lacustrine soils contain high amounts of inorganic carbonates. Measurements carried out in various soils in Greece have shown amounts of inorganic carbonates ranging from 17.2 to 67.5%. Organic soils contain lower amounts of carbonates. A recent soil survey showed that soils with strong or light reaction to hydrochloric acid cover 54.9 and 35.7%, respectively. Some organic soils are free of carbonates in the upper layers covering about 11.2%.

Even though lacustrine and organic soils are very poorly drained, soluble salts of sodium or potassium chloride are in low concentrations since the ground water is usually of low electrical conductivity (lower than 4 ds/m). However, white spots in the soil surface are common, especially in the organic soils, during the dry period. The chemical analysis of this material has shown that the main salt is magnesium sulfate favored by the decomposition of organic materials releasing sulfates. Karyotis et al. (2000) have found, in the Filippi Histosols, quantities of extractable heavy metals ranging from 7190 (Fe) to 5.1 (Cd) mg/g.

The biomass production potential of the soils of this group is one of the highest in the country, provided the soil moisture is controlled to levels that provide both the necessary oxygen and water levels to plant roots, and at the same time, it does not allow fast decomposition of the organic matter.

7.2 Management Groups

The recommended management of these is the agricultural development. Their suitability for farm crops is high, provided the water table depth is effectively controlled. They are not suitable for plant species sensitive to prolonged water saturation of the soil. The soils are unsuitable for plants such as perennial or annual winter crops.

Drainage networks have been constructed in the majority of organic and lacustrine soils of Greece, where the ground water level is mechanically controlled by bars at the outlet of the network. During the dry period, water level in the drainage ditches increases for keeping the water table in the level of about one meter from the soil surface to allow mechanization of cultivation and subirrigation of the growing crops. However, the amount of water from the existing rivers or springs supplying water to these areas is not adequate to satisfy the evapotranspiration needs of the growing crops resulting in lowering of the water table deeper than 2.5 m. Under these conditions, the following adverse consequences are raised: (a) crops cannot grow normally without the supply of additional irrigation water and (b) air porosity of the soil increases favoring oxygen diffusion resulting in organic material oxidation and degradation. Organic and lacustrine soils are mainly cultivated with annual spring crops. Some of these soils in connection with the surroundings water bodies are characterized as wetlands, protected by Natura 2000 or Ramsar Convention.

Ranges of average values of soil properties over the various bioclimatic zones of Greece and data of typical profile are given in the following Tables 6.3, 7.1, and 7.2.

Profile: **Calcaric gleyic Fluvisol (Aquic Xerofluvents) (P24-** Fig. 3.1)

Table 7.1 Ranges of mean values of clay content and chemical properties of soil horizons in profiles of soils formed in marls mainly in cropland areas (Derived from various soil survey studies of the NAGREF and Agricultural Univesrity of Athens)

Horizon	Clay%	pH	OM%	N%	C/N	Extr. P Ppm
Ap	26.7–70.0	7.7–8.2	1.6–37.3	0.78–3.41	8.2–12.5	5.3–155.9
C1	4.7–74.0	7.6–8.4	0.8–77.2	0.99–1.34	9.3–14.0	1.1–31.9
C2	2.7–80.0	7.6–8.4	0.1–62.5	0.15–4.50	10.1–16.1	1.0–27.9

Table 7.2 Morphological and analytical data (by: Kosmas et al. 2000, unpublished)

Horizon	Depth cm	Clay %	Silt %	Sand %	Gravel/stones	Structure	Org. mat. (%)	C/N %	pH
1Ap1	0–30	48.0	31.0	21.0	None	mfcr	11.6	12.5	7.6
1Ap2	30–60	54.0	31.0	15.0	None	mfcr	14.0	14.7	7.3
2C1	60–130	39.0	20.0	41.0	None	Svcpl	22.7	12.2	7.4
3C2	130–160	64.0	25.0	11.0	None	Svcpl	1.2	11.7	7.5

Horizon	Color Munsell	Exchangeable mg/Kg meq/100 gr			CEC cmol/Kg	BS %	Extractable P mg/kg	Total P mg/Kg
		Ca	Mg	K				
1Ap1	10YR 4/4	21,600	6075	156	68.1	100	34.2	
1Ap2	10YR 4/4	19,200	6464	195	58.0	100	13.1	
2C1	10YR 5/1	184,006	4860	39	48.2	100	4.3	
3C2	10YR 4/1	20,400	91,112	234	53.5	100	3.5	

Table 7.3 Analytical data of the above-described Typic Medisaprist soil profile (Yassoglou 1963)

Horizon	Depth cm	Organic matter %	Total N %	C/N	pH	CO_2 extractable ppm	
						P_2O_5	K_2O
Oa	0–80	34.2	1.27	15.7	7.8	0.27	13
Oa/Oi	80–120	74.7	1.71	25.4	6.8	0.27	13

Horizon	Me/100g soil					Carbonates %
	C.E.C	Exchangeable				
		K	Na	Mg	Ca	
Oa	78	0.4	0.8	9.2	67.6	8.0
Oa/Oi	115	0.4	2.3	19.4	73.9	traces

Region: Xiniada, Central Greece
Parent material: Lacustrine deposits
Physiography: nearly flat (slope 0–2%) Aspect: plain
Elevation: 450 m
Vegetation zone: Sub-Mediterranean
Land use: tomato

Profile: Sapric Histosol (Typic Medisaprists) (P25-Fig. 3.1)
Region: Philippi Marshes, East Macedonia
Parent material: Organic lake deposits
Physiography: nearly flat (slope 0–2%) Aspect: plain
Elevation 200 m
Vegetation zone: Sub-Mediterranean
Land use: Cultivated plants (Table 7.3)

Horizon	Depth cm	Description
Oa	0–88	Well-decomposed organic matter, black (10YR2/1) with blotches or discontinuous clay loam or silty clay loam mineral layers, deposited by the nearby stream; structure: granular: consistency: wet soft, dry hard; roots abundant; boundary to lower horizon: clear
Oa/Oi	80–120	Well-decomposed organic matter black (10YR 2/1). It contains blocks of fibrous peat of dark brown color (7.5 YR 3/2); roots abundant; boundary to lower horizon: gradual
Oa/Qig	+120	Well-decomposed organic matter black (10YR 2/1). It contains blocks of fibrous peat of dark brown color (7.5 YR 3/2); roots: abundant. Water saturated

8 Soils on Alluvial Deposits

The dominant soil-forming factors that strongly affect the properties, functions, and productivity of the alluvial soils are the intense stratification of their parent material, its textural configuration and composition, the age of the soils, which is related to the topographic location on the successive terraces and the climate.

Greek alluvial deposits form terraced landscapes of Holocene to Pleistocene—Pliocene age. In this group are also included Quaternary lacustrine formations. Thus, the soils developed on these landscapes constitute topo-chrono sequences, topography.

A very important property of the alluvial soils of Greece is their hydromorphy as expressed by their drainage classes, which corresponds to and distribution of soil moisture throughout the year. Drainage classes used in soil characterization and mapping are the following (Yassoglou 1995):

A: Dry excessively drained profile
B: Moderately dry well-drained profile
C: Moderately wet moderately well-drained profile
D: Wet imperfectly drained profile
E: Very wet poorly drained profile
F: Badly drained profile with permanently saturated sections of the profile at a depth of 50 cm and below from the surface of the soil.
G: Very badly drained profile with a permanent water table at depths shallower than 50 cm from the surface of the soil.

The above hydromorphic classes are applied to soils of the subhumid climatic zone of Greece. Soils in classes E and F, have temporary saturated sections in their profile during the raining season. Waters tables are also temporary characterized as perched water tables. Poorly drained soils are also found in the drier eastern provinces in soils with shallow phreatic water table.

Poorly drained and badly drained soils are usually found in fine textured (silty clay and clay) profiles of the subhumid zone.

The characterization of the hydromorphic class is determined on the basis of the morphological observations of the profiles in the field. The characteristic used are as follows: the active presence or absence at specific depths in the profile of Fe–Mn mottles and concretions, the chroma of Munsell color charts, and the presence of the glei horizon with bluish-neutral color (N/0 Munsell).

All the drainage classes can be found in the various landscapes of the Holocene Alluvial soils. Whereas, on the Pleistocene–Pliocene deposits, the classes usually predominate are the A, B, C, and D, depending on the climatic zone and the topography.

8.1 Soils on Holocene Alluviums

As it was referred above (chapter A), soils on Holocene alluviums are the most productive soils of the country. They are divided in those of age older than 2000 years and the recent alluvium which are younger than 2000 years. The main soil association found on Holocene alluvium parent material in Greece is Fluvisol with dominant soil typological unit (STU) according to FAO system Calcaric Fluvisols or dominant soil taxonomic unit according to Soil Taxonomy System Typic Xerofluvents. The associated STU FAO or taxonomic units are presented in the soil association map (Fig. 3.1). The quality of the soils included in these units is very high in almost all cases except for those with poor drainage (fluvic calcaric gleysol, including gleic solonchak or typic fluvaquents including Aquollic Salorthid). Vulnerability of these soils to desertification is considered low to medium (Table 3.2).

N. Yassoglou et al., *The Soils of Greece*, World Soils Book Series,
DOI 10.1007/978-3-319-53334-6_8

8.1.1 General Soil Properties of Fluvisols

A basic property of these soils is their fine texture and friability unless they have undergone packing due to poor management. Because of their friability, soil-forming processes on these materials are quick compared to hard autochthonous parent materials. Water movement into the deeper layers of the soil profile is easier, while soil's flora and fauna growth is favored. On the other hand, due to their fine texture these materials have high chemically active specific surface. These conditions favor chemical and physical weathering of minerals, solids and solution movement, and creation of organic matter which are important soil genesis processes.

Soil texture very strongly affects soil-forming processes. In extreme cases such as in very coarse or very fine sediments, this influence is so strong that climate influence on soil formation is minimized and climatic soils are developed. In the soils of flood plains, their texture is coarse close to the river beds becoming finer as the distance increases. In the flood basins, texture is becoming more homogeneous and finer. In a study of the area of Pinios River in west Peloponnese, it was found that the texture of the surface horizons was similar to that of the subsurface layers when it was coarse (sand or sandy loam) or fine (clay or clay loam), whereas the overlaying layers on subsurface layers with medium texture (loam) had very different textures from sandy loam to clay loam (Yassoglou et al. unpublished data). This strong stratification of the overlaying layers to loamy deposits is attributed to the different velocities of the flood water.

Mineralogical composition of the texture fractions depends on the origin of the deposits and the kind of rocks from which they come from. Sand and silt fractions originating from granite rocks contain mainly quartz, feldspars, and micas. Whither the fractions come from limestone, the minerals are rich in calcium carbonate but those coming from basic rocks contain olivine, pyroxene, and amphiboles. On the other hand, the composition of clay fractions depends mainly on the kind of soils in the area and indirectly on the rocks dominating in the area. Soils with advanced genesis contain clay rich in iron oxides and kaolinite. Less developed soils contain minerals with intermediate stability such as montmorillonite, vermiculite, and chlorite. Sand and silt fractions originating from the erosion of well-developed soils contain mainly quartz. Soils formed on basic rocks and marls contain clay rich in montmorillonite. Consequently, alluvial deposits coming from the erosion of these clays give rise to the formation of vertic characteristics (vertisols or vertic subgroups of Inceptisols). Plenty of these soils are met in Evros plain and Thessaly plain. Calcium carbonate dominates in Greek alluvial deposits. This fact plays an important role in soil formation processes since calcium carbonate increases soil pH and causes aggregation of clay colloids slowing the acidification of the system and clay movement. Argillic horizons can be formed on the older than 2000 years upper terrace of the landscape in Haploxeralf profiles, only after leaching of calcium carbonate from the upper parts of the soil solum. Thus, the regular course is the formation firstly of cambic horizons which characterize the order Inceptisol and later argillic horizons which is the diagnostic characteristic of the order Alfisol. (See also Chap. 2, Parent Material Affected Soil Properties)

In arid or semiarid areas, calcic horizons are formed. In general, deposits originating from basic igneous rocks, micaceous schists, limestone, and flysch give rise to fertile and very productive soils.

It is well known that the presence and movement of water in a soil system seriously affects most of the soil-forming processes. So factors associated with water presence and movement in the soil system are of significant importance. These factors are climate (rainfall, evaporation), relief, and texture of parent material. The relationship between these three factors determines the drainage and leaching of alluvial soils. Prolonged water saturation and seasonal alteration between water logging and drainage significantly affect soil chemical and morphological properties. Water saturation status of the soils has profound effects on the supply of oxygen to the soil, and in turn to the oxidation state of the elements such as iron, manganese, and sulfur which have significant influence on their solubility and color, giving rise to the formation of brown, gray, blue, and black mottles and fine Fe–Mn concretions that are very often met in hydromorphic soils. In addition, water saturation affects redo processes involving production or consumption of hydrogen cations that strongly affects soil's pH.

In areas with arid or semiarid climate, alluvial soils are dry, at oxidative conditions without mottles, i.e., with excessive or good drainage. Only in areas with appropriate relief and hydro-geological conditions where the phreatic ground water table is shallow, poor drained soils can be formed having horizons with glei properties, i.e., with signs of chemical reduction arising from saturation with groundwater which are the presence of marked concentrations of dissolved Fe (II) and gray color soil matrix with or without black or brownish to reddish mottles of Mn and Fe oxides. In subhumid or humid areas, drainage of alluvial soils is determined by texture and relief. Clayey soils with gentle slope have usually imperfect or bad or poor drainage (D or E at Yassoglou 1972), while sandy soils have excessive or good drainage (A or B). Carbonate salts' translocation takes place down the soil profile, at the lower parts of the flood plains, in the humid and sub-humid zones. This process is facilitated by the sufficient amounts of leaching water. In this way, cambic horizons right after the surface horizons or calcic horizons at deeper parts are developed so that soil systems Inceptisol are to be formed. Under these conditions with alterations of oxidative and reductive processes, usually mottles and concretions of iron and manganese are formed. Such soils, i.e., Inceptisols with gleic characteristics are formed in western Greece. In contrast to the dry eastern zones under similar conditions, soil horizons are not formed and the soils belong to the suborder Xerorthents. At the upper parts of alluvial deposits where the deposition rate is very slow or zero, after some thousands years Inceptisols first and Haploxeralfs later with argillic horizons are developed. In the area between the current bed of the water flow and the floodplains where the deposition rate is quite fast and there is no much water to enhance soil genesis processes, no horizons are formed and the soil systems remain Entisols. In western Greece where the climate is humid, the usual sequence of soil systems between active and inactive alluvial elevations is Entisols with good or poor drainage, Inceptisols with good or poor drainage, and Alfisols with good drainage (Yassoglou unpublished data).

Alluvial deposits geologically are recent formations but from soil genesis point of view, they have a quite high age graduation permitting significant differentiation of soil systems. Alluvial soils in Greece belong to Quaternary era and more specifically to the last part of the Pleistocene and Holocene. It means that the range of the ages of alluvial soils varies significantly. So, time as a soil genesis factor in some cases plays a significant role. This is obvious in Greek alluvial plains where there is a close relationship between relief and age of deposits which gives rise to the development of topo-chrono sequences. Thus, alluvial plains consisted of a series of terraces; soil series with different ages and evolution are met. Starting from the current alluvial flood plain of the water flow and going to higher elevations up to about 30 m from the starting point, we meet the following soil systems. As explained earlier, Entisols formed close to water to levies and banks of rivers and steams. Whether in wet places, away from them, cambic horizons of Inceptisols are eveloped. Inceptisols also are found in higher elevations which in areas of southeastern Greece where the climate is dry have calcic horizons.

8.2 Soils on Pleistocene–Pliocene Deposits

At higher elevations and up to about 30 m from the sea or inland flood plain, Alfisols with well-developed argillic horizons and different colors are met. In dry areas like in southwestern Greece, the color is intense red or reddish brown reflecting the good soil drainage conditions. In some cases in the same areas, a calcic horizon is formed. This time series is found in several coastal areas of eastern Greece and the Thessaly plain. In contrast to western Greece in areas with humid climate, the existence of argillic horizons creates poor drainage conditions, leading to the development of Fe and Mn mottles and concretions. Soil color is brownish yellow or reddish yellow. In some places where the weathering of clay minerals is intense, E horizons with very low clay content are developed and the soils are classified as Planosols (Palexeralfs). These soils are characterized by an abrupt change in clay content between the argillic horizon and the overlying horizons. In these soils, there is an abundance of Fe and Mn concretions. These soils are met in western Greece where the climate is characterized by high annual rainfall (up to 1200 mm). However, base saturation of these soils usually is greater than 50% and usually 60–80%. Sometimes these soils have frangipanes. Important role in soil formation processes seems to play natural vegetation such as quercus and erica which through the release their roots of organic substances enhance minerals weathering and soil leaching.

In areas with elevation greater than 50 m from the lower part of the flood plains, the soils were developed on stoniness parent material deposited during the Pleistocene period on tertiary formations. This material called *Rania* is met very often in Mediterranean countries. On these deposits, advanced developed Alfisols are found being rich in minerals, quartz and jasper, that are very persistent in weathering. Due to the fact that in clay fraction of these soils, kaolinite predominates and cation exchange capacity is very low.

On the upper most terraces of this land form, several soils, in the wetter sections of the country, have been formed in the subsurface horizons numerous coarse Fe–Mn concretions. On some of these terraces, some soils have acquired ultic characteristics due to extensive leaching.

An important group of soils on Pleistocene–Pliocene terraces are the Red Mediterranean Soils of Greece. They belong mostly to the great groups of Rhodoxeralfs, Palexeralfs, Xerochrepts, and Xerorthents. These allochthonous soils have likely inherited and retained their color from their parent materials, which were transported from original soil horizons formed on hard limestone. These soils retain their color only on sloping well-drained terrain in the thermo-Mediterranean zone. The soil profiles are frequently stratified and occasionally bisequal. Their original parent material was rich in iron-bearing minerals and bases. It contained small not deeply distributed organic matter (Yassoglou et al. 1997).

The properties of these soils have been acquired under the present climatic condition, since climate remained stable for the last 5000 years (Yassoglou and Nobeli 1972)

9 Specific Characteristics of Alluvial Soils

Soils on Holocene alluvium are classified as Fluvents (Fluvisols), as it was referred above these soils including the most productive agricultural soils covering about 13% of the total area of Greece (see Fig. 9.1). Fluvisols include the soil associations 19, 20 21, 22, 23, 24, 25 (Fig. 9.2) the distribution of which is shown in the Table 9.1. It is estimated that Fluvisols occupy about 45% of the Greek land that is used for agriculture. This fact is indicative of the importance of this soil order to the economy of the country. Fluvisols occupy about 45% of the area used for agriculture in Greece. This fact reveals the significance of this soil order for agricultural production of the country.

The ranges of the values of the basic soil properties of surface horizons of Fluvisols are shown in the Table 9.1.

In all the associations of Fluvisols, the dominant soil typological unit (STU) according to FAO soil classification system is Calcaric Fluvisol or according to Soil Taxonomy System (STS) Typic Xerofluvents (soil associations 19–25). A small area also is covered by soils developed on Holocene alluvium and older terraces included in the soil association 33 (see Fig. 7.5) with dominant Typological unit Calcaro-vertic Cambisol (FAO system) or Vertic Xerochrept (Soil Taxonomy system). A more detailed description of soil properties included in these associations which mainly are Fluvisols (FAO) or Entisols (Fluvents) (STS) follows.

9.1 Soils Included in Soil Association 19 (Calcareous Entisols, Inceptisols with Inclusions of Solonchak on Holocene Alluvium)

This soil association extends in all country occupying about 55% of the Fluvisols in Greece. Soils included in this association were formed on alluviums of coastal floodplains or inland floodplains. They are found in northern Greece (Thrace, Central and West Macedonia, western Greece, Thessaly, Attica, and Peloponnesus). The dominant Soil Taxonomic Unit as already referred is Typic Xerofluvents. In coastal flood plains, these soils are found at elevations up to about 20–30 m above sea levels or about 30 m higher above the water flow in inland flood plains.

These soils are deep, with medium texture depending on the distance from the water flow, with calcium carbonate in the whole soil profile or below surface layer, with organic matter content around 1.8% in the surface layers and well below in subsurface layers. Soil reaction is neutral or slightly alkaline (see Table 9.2).

Description of Soil profiles of Soil Association 19

Profile: Calcaric Fluvisol (Typic Xerofluvents) (P26-Fig. 3.1)
Region: Kouloura, Larisa
Parent material: Holocene Alluvium
Physiography: Flat (slope 0–2) Aspect: plain
Elevation: 2 m
Vegetation zone: Thermomediterranean
Land use: Corn (Table 9.4)

Description of Soil profiles of Soil Association 19

Profile: Calcaric Fluvisol (Typic Xerofluvents) (P27-Fig. 3.1)
Region: Piniada, Trikala
Parent material: Holocene Alluvium
Physiography: Flat (slope 0–2) Aspect: plain
Elevation: 100 m
Vegetation zone: Thermomediterranean
Land use: Wheat (Table 9.5)

Profile: Calcaric Fluvisol (Typic Xerofluvents) (P28-Fig. 3.1)
Region: Makrychori Karditsas
Parent Material: Holocene alluvium
Physiography: Slope 0–2% Aspect: plain
Elevation: 94 m
Vegetation zone: Thermomediterranean
Land use: Wheat (Table 9.6)

N. Yassoglou et al., *The Soils of Greece*, World Soils Book Series,
DOI 10.1007/978-3-319-53334-6_9

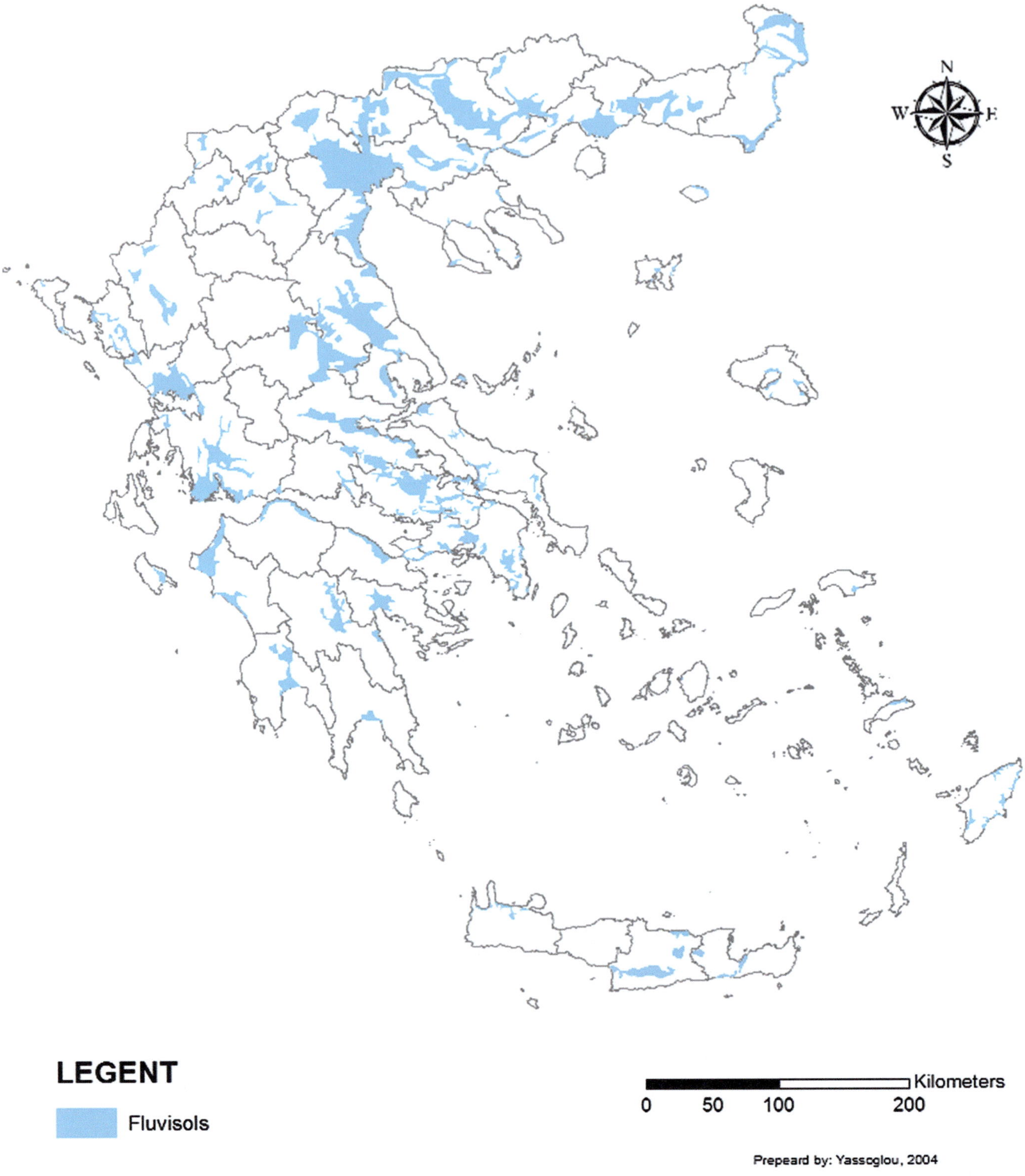

Fig. 9.1 Distribution of Fluvisols in Greece

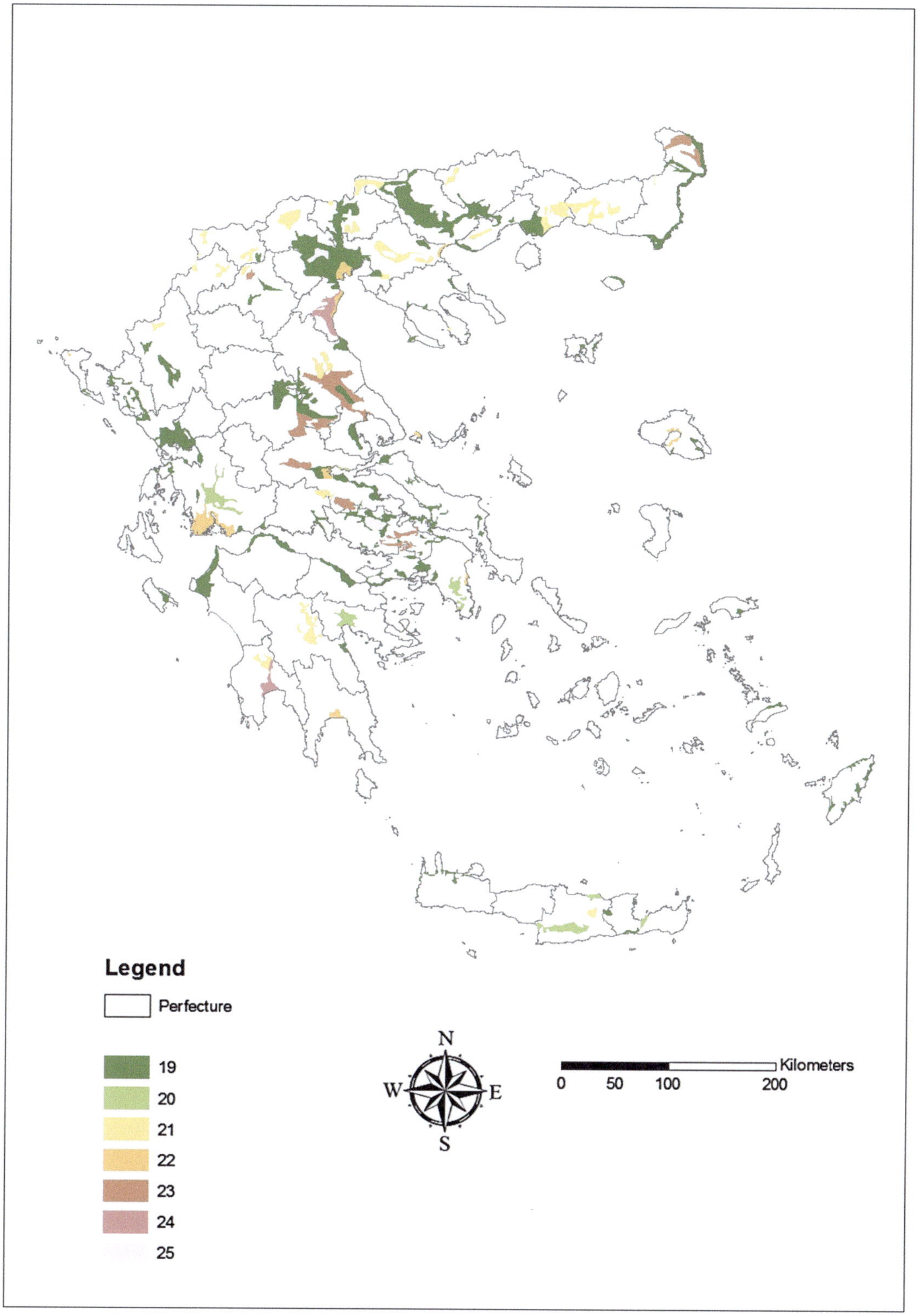

Fig. 9.2 Distribution of soil Associations of Fluvisols

Table 9.1 Area of the soil associations included in the soil order Fluvisols

Soil association	Area, km^2	% of the country area	% of the soil order Fluvisols
19	9966.309	7.55	55.5
20	1163.964	0.88	6.4
21	2891.800	2.19	16.1
22	967.676	0.73	5.4
23	2022.579	1.53	11.3
24	542.043	0.41	3.0
25	412.707	0.31	2.3
Sum	17967.080	13.62	100

Table 9.2 Ranges of mean values of the main physicochemical properties of surface horizons of Fluvisols (ISMC 1989a, 1989b, 1990a, 1990b, 1991a, 1991b, 1991c)

Property	Minimum	Soil association	Maximum	Soil association	Descending order between associations
Sand,%	21.40	20	62.29	21	20 < 22 < 24 < 19 < 23 < 21
Silt,%	21.78	21	37.36	20	21 < 23 < 19 < 22 < 24 < 20
Clay,%	15.90	21	41.50	20	21 < 24 < 22 < 19 < 23 < 20
pH	7.09	19	8.07	20	19 < 21 < 24 < 22 < 20 < 23
$CaCO_3$, %	3.68	21	23.15	20	21 < 22 < 19 < 24 < 23 < 20
O.M., %	1.67	23	2.83	24	21 < 23 < 22 < 19 < 20 < 24
CEC, cmol(+)/kg	15.71	21	33.81	22	21 < 20 < 23 < 19 < 24 < 22
EC, mmhos/cm	14.13	19	14.13	22	19 < 23 < 22
P Olsen, mg/kg	6.45	24	22.72	21	24 < 22 < 19 < 20 < 23 < 21
Kexch., cmol(+)/kg	0.31	20	2.42	23	20 < 22 < 21 < 19 < 23

Table 9.3 Ranges of values of selected physicochemical properties of surface horizons of soil association 19 (Fluvisols, data of the National Agricultural Research Foundation of Greece data base)

Value	Soil properties											
	Sand[a]	Silt[a]	Clay[a]	pH[b]	$CaCO_3^a$	CEC[c]	O.M.[a]	Ntot[a]	EC[d]	POlsen[e]	Kexch[c]	Mgexc[c]
Mean	32.01	30.36	37.90	7.09	7.95	29.04	2.10	0.10	4.40	13.33	1.70	4.88
Min	5.00	12.00	6.90	4.10	1.30	8.09	0.34	10.06	0.44	2.30	0.08	0.70
Max	79.70	56.00	84.00	8.80	29.82	52.60	22.00	0.18	12.20	28.00	12.00	20.10
Median	31.00	29.00	36.00	7.30	7.30	28.30	1.79	0.08	0.67	14.30	1.40	1.60
N[f]	151	151	151	22	69	32	141	90	50	70	37	16

[a]%, [b]Soil/water 1:5, [c]cmol(+)/kg, [d]mmhos/cm, [e]mg/kg soil, [f]Number of observations

9.2 Soils Included in Soil Association 20 (Calcareous Inceptisols, Entisols, and Rhodoxerlafs on Holocene Alluvium and Holocene Terraces)

This soil association covers about 1164 km^2 which corresponds to a percentage of the total country area 0.88 and 6.4% of the Fluvisols (Table 7.3). The quality of these soils is very high, and the vulnerability to desertification is quite low. They are suitable for agronomic use with very strong restriction for non-agronomic uses (Table 9.3). A representative soil profile is presented in Table 9.7 and Fig. 7.2.

Soil profiles of Soil Association 20

Profile: Calcaric Cambisol (Typic Xerochrepts) (P29-Fig. 3.1)

Table 9.4 Morphological and analytical data (ISMC 1997)

Horizon	Depth, cm	Clay, %	Silt, %	Sand, %	Texture class	Structure	Color Munsell	Boundary	$CaCO_3$, %
Ap1	0–16	22	42	36	L	1mabk	10YR5/3	Ds	6.1
Ap2	17–31	20	37	43	L	1mabk	10YR4.5/3	Cw	6.1
C1	32–47	10	31	59	SL	Structurless	10YR5.5/3.5	Cs	8.8
C2	48–58	8	14	78	SL/LS	Structurless	10YR5.5/3.5	Cs	8.6
2C3	59–68	14	33	53	SL	1mabk	10YR3.5/3	Cs	8.2
3C4	69–92	32	47	21	CL	2mabk	10YR4/3	Cw	5.7
3C5	93–114	22	43	35	L	Structurless	10YR5.5/3.5	–	8.4
4C6	115–150	10	35	55	SL	Structurless	10YR3.5/3	–	7.6
5C7	151–180	34	49	17	SiCL	Structurless	10YR3.5/3		4.6

Horizon	Depth, cm	pH 1:5	Org. matter, %	Exchangeable cations, cmol/kg			Cation exchange capacity, cmol/kg	EC, mmhos/cm
				Na^+	K^+	ESP, %		
Ap1	0–16	8.0	1.27	0.15	0.21	0.90	16.5	<3
Ap2	17–31	8.1	1.25	0.10	0.19	0.62	15.9	<3
C1	32–47	8.4	0.33	0.10	0.10	0.82	12.1	<3
C2	48–58	8.4	0.23	0.10	0.08	1.21	8.2	<3
C3	59–68	8.3	0.40	0.10	0.10	0.79	12.6	<3
2C4	69–92	8.3	1.24	0.15	0.19	0.49	31.3	<3
3C5	93–114	8.2	0.67	0.15	0.13	1.01	14.8	<3
3C6	115–150	8.3	0.27	0.10	0.06	1.13	8.8	<3
3C7	151–180	8.3	0.60	0.15	0.21	0.60	24.7	<3

Table 9.5 Morphological and analytical data (ISMC 1991b)

Horizon	Depth, cm	Clay, %	Silt, %	Sand, %	Texture class	Structure	Color Munsell	Boundary	$CaCO_3$, %
Ap	0–24	27	49	24	L	Massive	10YR5/4	g	8.61
A1	24–52	29	50	20	L/SL	1mabk	10YR4/4	g	8.40
BC1	52–98	37	25	38	CL	2mabk	10YR3/3	g	1.63
BC2	98–135	25	13	62	SCL	1msbk	10YR4/4		6.3

Horizon	Depth, cm	pH 1:5	Org. matter, %
Ap	0–24	8.0	1.17
A1	24–52	8.2	0.98
BC1	52–98	8.1	1.03
BC2	98–135	8.1	0.40

Region: Polyfyto Kozanis
Parent material: Holocene alluvium
Physiography: Slope 2–6% (Aspect: Plain)

Elevation: 315 m
Vegetation zone: Thermomediterranean
Land use: Wheat (Fig. 9.3)

Table 9.6 Morphological and analytical data (ISMC 1991a)

Horizon	Depth, cm	Clay, %	Silt, %	Sand, %	Texture class	Structure	Color Munsell	Boundary	$CaCO_3$, %
Ap	0–25	18	56	26	SiL	1fgr	10YR5/3	ds	0
A	25–50	14	56	30	SiL	1fgr	10YR5/3	as	Traces
2C1	50–73	18	56	26	SiL	1msbk	10YR5/4	as	Traces
3C2	73–105	8	26	65	SL	Massive	10YR5/4	as	0
3C3	105–138	12	44	44	L	Massive	10YR5/4	ds	0

Horizon	Depth, cm	pH 1:5	Org. matter, %	CEC, cmol/kg
Ap	0–25	7.1	1.65	27.6
A	25–50	7.5	0.45	22.8
2C1	50–73	7.7	0.56	26.4
3C2	73–105	7.8	0.14	20.7
3C3	105–138	7.8	0.24	–

[a]Fe concretions below 50 cm

Table 9.7 Ranges of values of selected physicochemical properties of surface horizons of soil association 20 (Fluvisols, data of the National Agricultural Research Foundation of Greece data base)

Value	Selected properties									
	Sand[a]	Silt[a]	Clay[a]	pH[b]	$CaCO_3^a$	CEC[c]	O.M.[a]	Ntot[a]	POlsen[d]	Kexch[c]
Mean	21.14	37.36	41.50	8.07	23.15	23.74	2.28	0.12	13.33	0.32
Min	2.60	26.60	20.40	7.40	1.10	12.10	1.20	0.02	2.30	0.16
Max	53.00	66.00	64.00	8.50	39.50	46.00	5.40	0.32	28.00	0.56
Median	17.10	33.70	40.70	8.10	22.90	24.00	2.15	0.12	14.30	0.32
n	41	41	41	22	22	17	22	22	70	22

[a]%, [b]Soil/water 1:5, [c]cmol(+)/kg, [d]mg/kg soil, [e]Number of observations

9.3 Soils Included in Soil Association 22 (Calcareous Aquents, Aquepts, and Solonchak on Holocene Alluvium)

This soil association is similar to soil association 20. It occupies an area of about 968 km^2 corresponding 5.4% of the total area covered by Fluvisols. The quality of soils which are suitable for agricultural use is very high and have low vulnerability to desertification These soils are not suggested for non-agricultural use (Tables 9.3 and 7.3). Morphological and analytical data of these soils are presented in Table 9.8.

Profile: Clacaric Fluvisol (Aquic Xerofluvents)
Region: Karditsa
Parent Material: Holocence Alluvium
Physiography: Slope 0–2%
Elevation: 110 m
Vegetation zone: Thermomediterranean
Land use: Annual crops (Table 9.9)

9.4 Soils Included in Soil Association 23 (Calcareous Chromoxerets, Pelloxererts, Haploxeralfs, and Xerochrepts on Holocene Alluvium and Quaternary Terraces)

The associated typological soil units of this association are Chromic Haploxerert, Typic Haploxerert, Typic Haploxeralf, and Calcixerollic Haploxeralf. These soils are xtended in an area of about 2022 km^2 covering a percentage of 11.3 of the total area of Fluvisols. Quality of these soils is vey high and are very suitable for agricultural use having low vulrenability to desertification (Tables 3.2 and 7.3). Representative morphological and analytical data of these soils are presented in Table 9.8 and Fig. 7.3).

Soil profiles of soil association 23
Profile: Haplic Luvisol (Typic Haploxeralfs) (P31-Fig. 4.3)
Region: Ryakia Pierias, Macedonia

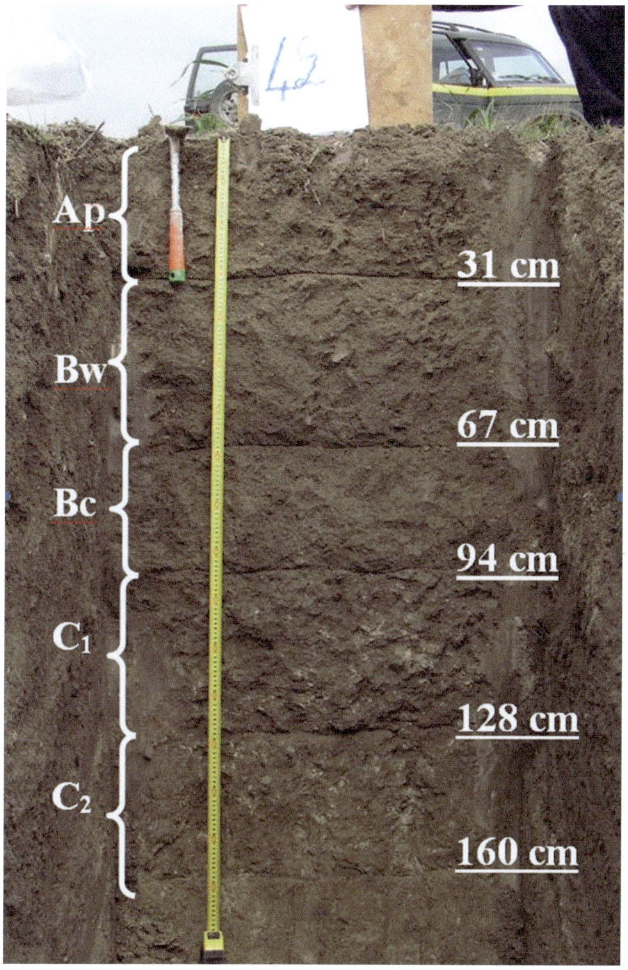

Fig. 9.3 A Calcaric Cambisol (Typic Xerochrepts) (Polyfyto Kozani, Photo F. Papadopoulos) (P29-Fig. 4.3)

Parent material: Quaternary Terraces
Physiography: Gently sloping terrace (slope 2–6%) Aspect: plain
Elevation: 300 m (aproxm.)
Vegetation zone: Mesomediterranean
Land use: Annual crops (Fig. 9.4 and Table 9.10)

Profile: Calcic Vertisol (Typic Calcixererts) (P32-Fig. 4.3)
Region: Neo Ikonio Karditsa
Parent Material: Quaternary terrace
Physiography: Nearly flat terrace (Slope: 0–2%) Aspect: Plain
Elevation: 300 m (aproxm)
Vegetation zone: Mesomediterranean
Land use: Annual crops (Fig. 9.5 and Table 9.11)

9.5 Soils Included in Soil Association 24 (Calcareous Aquents, Aquepts with Solonchak on Recent Alluvium)

The associated typological soil units of this association are Fluvicalcaric gleysol and gleic Solonchak (Table 3.2). These soils are characterized by a poor drainage since the ground water table is shallow, and during the winter periods, its depth is 25–50 cm below soil surface. These soils have layers with different properties such as texure, organic matter content, and color. In order these soils to become productive, it is needed to be drained especially when they are going to be cultivated with trees or perennial crops.

Soil profiles of Soil Association 24
Profile: Mollic Solonchak (Aquollic Salorthid) (P34-Fig. 4.3)
Region: Aiginio Pierias
Elevation <100 m

Table 9.8 Morphological and analytical data (ISST Institute of Soil Science of Thessaloniki 2006)

Horizon	Depth, cm	Clay, %	Silt, %	Sand, %	Texture class	Structure	Color Munsell	Boundary	$CaCO_3$, %
Ap	0–31	34.0	31.6	34.4	CL	m	10YR5/2	g	33
Bw	32–67	36.4	29.2	34.4	CL	3msbk	10YR4/2	g	28.6
Bw/Bc	68–94	36.4	31.2	32.4	CL	2msbk	10YR5/2	g	38.7
C1	95–128	38.4	27.2	34.4	CL	3msbk	10YR5/3	g	7.0
C2	129–160	38.4	33.2	28.4	CL	2msbk	2.5YR5/3	g	7.9

Horizon	Depth, cm	pH 1:5	Org. matter, %	CEC, cmol/kg	Exch. K, cmol/kg	Exch. Mg, cmol/kg
Ap	0–31	8.04	1.35	32.58	0.42	1.01
Bw	32–67	8.15	1.15	35.84	0.42	1.38
Bc	68–94	7.62	0.81	30.95	0.30	0.65
C1	95–128	8.23	0.60	26.61	0.24	0.58
C2	129–160	8.09	0.87	24.98	0.24	0.69

Table 9.9 Morphological and analytical data (ISMC 1991a)

Horizon	Depth, cm	Clay, %	Silt, %	Sand, %	Texture class	Structure	Color Munsell	Boundary	$CaCO_3$, %
Ap	0–27	37	51	12	SiCL	1msbk	10YR5/2	s	0
C1	28–50	27	63	10	SiL	Massive	2.5YR4/4	Gs	0
C2	51–72	31	55	14	SiCL	Massive	2.5YR5/4	As	19.4
2C3	73–81	9	32	59	SL	Massive	5YR4.5/2	as	0.8
2C4	82–100	15	47	38	L	Massive	2.5YR5/2		20.5

Horizon	Depth, cm	Org. matter, %
Ap	0–27	3.1
C1	28–50	0.7
C2	51–72	0.3
2C3	73–81	0.6
2C4	82–100	0.2

Fig. 9.4 A Haplic Luvisol (Typic Palexeralf) (Agia Sofia Larissa, Photo S. Floras) (P30-Fig. 4.3)

Parent Material: Holocene alluvium
Physiography: alluvial plain (Slope 0–2%) Aspect: Plain
Elevation: 300 m (aproxm)
Vegetation zone: Mesomediterranean
Land use: Annual crops (Table 9.12)

9.6 Management Groups

Due to their formation, alluvial soils are characterized by heterogeneity of their profiles usually having layers with different properties. Depending on the prevailing conditions during the deposition of the parent materials, their texture and therefore almost all of their properties are related with it. The high heterogeneity of the profiles of the alluvial soils results in the creation of a big number of soil mapping units and soil series, so that it is difficult to utilize them in the agricultural practice. Therefore, it is necessary this big number of cartographic units to be grouped in a smaller number of units called *management groups*, based on general properties which seriously affect soil management. These properties are soil classification, soil drainage, and texture (1). According to the system applied in Greece (Yassoglou 1995), the following symbols are used to group cartographic units to management groups of Fluvisols and especially of Fluvents and Fluventic Xerochrepts:.

Geographic Location: First letters of the location's name

Taxonomic category	Fluvents—Fluventic Xerochrepts	EF + If
Drainage	Very well to well drained	A, B
	Imperfectly drained	C, D
	Poorly drained	E
	Very poorly drained	F, G

(continued)

Table 9.10 Morphological and analytical data (ISMC 1995)

Horizon	Depth, cm	Clay, %	Silt, %	Sand, %	Texture class	Structure	Color Munsell	Boundary	$CaCO_3$, %
Ap	0–17	46.8	40.0	13.2	SiC	2mgr	10YR3/3	cs	0
B1	18–43	42.4	42.0	15.6	SiC	2mpr	10YR3/3	vas	0
Bt1	44–75	44.8	43.0	12.2	SiC	3cpr	7.5YR4/4	as	0
Bt2	76–90	51.0	38.0	11.0	C	3cpr	7.5YR4/2	as	0
B2	91–110	37.0	46.0	17.0	SiCL	2mpr	7.5YR5/4	gs	0
C	111+	31.2	44.0	24.8	CL	Massive	2.5YR5/4		Traces

Horizon	Depth, cm	pH sat. past	Org. matter, %	CEC, meq/100 g	EC, mmhos/cm
Ap	0–17	6.47	2.1	53.4	0.31
B1	18–43	6.93	1.5	55.4	0.44
Bt1	44–75	6.00	0.7	60.8	0.53
Bt2	76–90	5.80	0.4	57.0	1.00
B2	91–110	6.30	0.3	57.0	1.36
C	111+	7.60	0.3	43.5	0.48

Fig. 9.5 A Calcic Vertisol (Typic Calcixerert) (Melissohori, Larissa, photo S. Floras) (P33-Fig. 4.3)

Taxonomic category	Fluvents—Fluventic Xerochrepts	EF + If
Texture	Medium (FSL, L, SiL)	1
	Slightly fine (SCL, LS, S, Si)	2
	Fine (SC, SiC, C)	3
	Coarse or silty (S, LS, or Si)	4

This categorization may be differentiated locally according to special properties of soils of the area such as soil depth, slope, existence of stoniness, and soil problems. A short description of the most important management groups in the area of the Fluvisols (Fluvents plus Fluventic Inceptisols) is presented below:

9.6.1 Thessaly

Thessaly is considered the most important agricultural area in the country. It covers about 14,000 km^2 corresponding to 11% of the total country area from which about 57% is agricultural land, where mainly cereals, cotton, corn, and in a less extent horticultural crops and grapes are grown. The region of Thessaly administratively is divided in four prefectures (Larissa, Magnesia, Karditsa, and Trikala). On the basis of its physiography (relief, geology, and climate) Thessaly is divided in two parts i.e., East Thessaly, including the areas of the prefectures Larissa and Magnesia and West Thessaly, including the prefectures of Karditsa and Trikala. These two parts differ substantially in their climate. In East Thessaly, climate is characterized according to Koeppen and Geiger as mesothermic Mediterranean with mean temperature of the coldest month >−3 °C, mean temperature of the warmest month >10 °C, and mean annual rainfall 430 mm. The climate of West Thessaly is characterized as warm temperate with mean temperature of the warmest month −3 to 8 °C, mean temperature of the warmest month 22 °C, and mean annual rainfall 740 mm.

Table 9.11 Morphological and analytical data (ISMC 1991a)

Horizon	Depth, cm	Clay, %	Silt, %	Sand, %	Texture class	Structure	Color Munsell	Boundary	$CaCO_3$, %
Ap	0–15	50	31	19	C	2fsbk	10YR5/2	d	0
A1	15–46	54	27	19	C	2msbk	10YR4/2	d	0
A2	46–107	58	25	17	C	3vfsbk	10YR4/2	a	5.3
Ck1	107–137	38	21	41	CL	2msbk	10YR5/6	a	30.7
C	137–157	38	28	34	CL	2msbk	10YR5/6		25.8

Horizon	Depth, cm	pH 1:5	Org. matter, %	CEC, meq/100 g
Ap	0–15	6.6	1.34	33.0
A1	15–46	6.6	1.07	28.0
A2	46–107	8.4	0.13	23.6
Ck1	107–137	9.0	0.07	20.0
C	137–157	8.8	0.13	24.0

Table 9.12 Morphological and analytical data (ISMC 1995)

Horizon	Depth, cm	Clay, %	Silt, %	Sand, %	Texture class	Structure	Color Munsell	Boundary	$CaCO_3$, %
Apz	0–24	33	53	14	SiCL	2mgr	2.5Y4.5/4.5	cs	14.2
C1z	25–40	29	59	12	SiCL	Massive	2.54/4	ds	14.8
C2z	41–52	23	61	16	SiL	Massive	5Y4/1–2.54/$	ds	12.4
2C3z	53–61	14	81	5	SiL	Massive	10YR5/6	ds	10.0
2C4z	62–80	12	80	8	SiL/Si	Massive	10YR5/4	ds	13.7
3C5z	81–99	34	40	26	CL	Massive	10YR4/3–5Y4/1	ds	12.1
4Cz6	100–110	46	43	11	SiC	Massive	2.5Y4/4		11.9

Horizon	Depth, cm	satur. past	Org. matter, %	CEC, cmol/kg	ESP,%	EC, mmhos/cm
Aps	0–24	7.8	1.4	20.3	21.5	16.30
Cz1	25–40	7.9	1.2	19.7	32.6	25.80
Cz2	41–52	7.8	1.1	16.3	4.8	55.50
2Cz3	53–61	7.7	1.4	23.3	38.4	69.00
2Cz4	62–80	7.5	1.1	20.0	26.4	67.00
3Cz5	81–99	7.7	1.0	27.4	21.4	60.75
4Cz6	100–110	7.7	1.6	30.4	34.5	52.00

9.6.1.1 Management Groups of East Thessaly (ISMC 1989a, 1989b, 1990a, 1990b)

The main management groups of this region and their basic characteristics are as follows (for the meaning of symbols of management groups see appendix).

Very well-drained, medium-textured Fluvents of Larissa area (LEfA1)

Deep soils with medium to coarse texture, very well drained, calcareous with weak to moderate subangular blocky or weak fine granular structure and good workability. Bulk density ranges between 1.65 and 1.75 g cm^{-3}. Their organic matter content is low around 1.5%, and they also are poor in nitrogen content. This management group is the largest in the area and most of the crops may be grown on that without serious limitations. The fertility of these soils is medium. However, usually they are rich in potassium but poor in phosphorus. In general, these soils are productive. The soils of this management group are suitable for all the crops grown in the area both non-irrigated like cereals (wheat, barley, oat, vetch) and irrigated such as corn, cotton, sugar beets, and tree crops. All types of fertilizers are appropriate, required in relatively high amounts of nitrogen and low amounts of potassium, especially in non-irrigated fields. Cultivation of soils does not present limitations.

Very well to well-drained, medium fine-textured Fluvents of Larissa area [LEf(A, B)2]
Deep soils with medium fine texture, very well to well-drained, calcareous with moderate granular structure in the surface layers and weaker in the deeper layers. Their workability is quite good. They contain low organic matter (around 1.7%), and they are poor in nitrogen content. The median value of bulk density is 1.40 g cm^{-3}. This management group is smaller than the previous and supports similar crops i.e., cereals, barley, oat, vetch (non-irrigated), corn, and cotton without any important limitation. The fertility of these soils is good, they are rich in potassium and phosphorus probably due to their fertilization In general, these soils are quite productive. All types of fertilizers are appropriate, required in relatively low amounts of nitrogen, potassium, and phosphorus, especially in non-irrigated fields.

Very well to well-drained, fine-textured Fluvents of Larissa area [L(Ef + If)(A, B)3]
Deep soils of fine texture, very well to well drained, calcareous in the whole profile or in the deeper layers only, with calcic horizons in some cases in a depth of around 1.0 m. Soil structure is moderate granular in the surface layers and weaker in the deeper layers. In the subsurface layers, their structure is mostly moderate subangular or angular blocky becoming massive in the deeper layers. Due to their fine texture, their workability is not so good and requires appropriate management especially with respect to their moisture content. They contain about 1.90% organic matter. Bulk density ranges between 1.47–1.49 g cm^{-3}. This management group is smaller than the two previous and supports crops with similar significance i.e., cereals, barley, vetch (non-irrigated), corn, and cotton (irrigated) with limitations related to their fine texture. Cation exchange capacity of the soils is quite high ranging between 20 and 35 cmol(+) kg^{-1}. Bulk density varies around 1.47 g cm^{-3}. The fertility of these soils is high, they are rich in potassium (from 0.57 to 1.89 cmol(+) kg^{-1} but low in phosphorus. In general, these soils are very productive and may support all the crops grown in the area except some tree crops sensitive to calcium carbonate. All types of fertilizers are appropriate required in relatively low amount of potassium but higher amount of phosphorus especially in irrigated fields. Special attention is required in soil management in relation to their moisture content.

Very well-drained, coarse-textured Fluvents of Larissa area (LEfA4)
Deep soils very well-drained of coarse-texture calcareous (calcium carbonate content from 2–15%) with weak structure in the surface layer or structurless in the deeper layers with medium infiltration rate (1–6 cm h^{-1}) and very good workability. Soil pH is neutral to slightly alkaline, organic matter content is low around 1.25% and potassium and phosphorus are also low. However, under appropriate management and fertilization, these soils may be very productive. These soils are suitable for most of the crops grown in the area such as rain feed cereals and irrigated vegetables or tree crops. The area covered by this management group is small compared to the previously described but it is important for the local communities.

Moderately drained, medium fine-textured Fluvents of Larissa area (LEfC2)
Deep soils, of medium fine texture, moderately drained with calcium carbonate usually in the subsurface layers with weak to moderate structure in the surface horizon and weak up to moderate subangular or angular blocky structure in the deeper layers with low to medium infiltration rate (0.3–6.0 cm h^{-1}). Soil workability is good but its accessibility is difficult in the wet period. Soil pH is slightly acidic to slightly alkaline, and organic matter content in average is 1.40%. These soils are suitable for late spring crops. Winter crops require drainage to remove surface waters by constructing shallow ditches.

Moderately drained, fine-textured Fluvents of Larissa area (LEfC3)
Deep soils, of fine texture, moderately drained with calcium carbonate in the whole profile with moderate structure in the surface horizon and weak up to moderate angular blocky structure in the deeper layers with medium low infiltration rate (0.5–2.0 cm h^{-1}). Soil workability due to its heavy texture and poor drainage but its accessibility is difficult in the wet periods. Soil pH is alkaline and organic matter content in average is 1.13%. These soils are suitable for late spring crops. Winter crops require drainage to remove surface waters by constructing shallow ditches. These soils are not suitable for tree crops unless their drainage will be adequately improved.

9.6.1.2 Management Groups of West Thessaly (ISMC 1991b, 1991c)

Very well to well-drained, medium-textured Fluvents and Fluventic Inceptisols of Karditsa and Trikala areas [(K–T) (Ef + If) (A, B)1]
Deep soils, of coarse to medium texture, very well to well-drained, mostly calcareous, with weak subangular to angular blocky structure, medium slow infiltration rate (0.5–2.0 cm h^{-1}), bulk density around 1.65 g cm^{-3} in the surface horizons being reduced in the deeper layers up to 1.42 g cm^{-3}, with good workability and accessibility. Soil pH is neutral (around 7.0) up to slightly alkaline (8.0) and organic matter content in average is 1.50%. Cation exchange capacity varies

between 17 and 28 cmol(+) kg^{-1} with a median value 23 cmol(+) kg^{-1}. Soil fertility is generally low with available P (P_{Olsen}) around 10 mg kg^{-1} soil, but rich available potassium which ranges between 1.0 and 4.0 cmol(+) kg^{-1} with a median value 1.50 cmol(+) kg^{-1}.

This management group is suitable for all the crops grown in the area without any serious limitations, especially for the most important crops of the area such as cereals and cotton. All the spring crops (corn, sugar beets) and the tree crops require irrigation in appropriate quantities according to plant kind needs. All the types of fertilizers are suitable requiring phosphorus in high quantities especially in irrigated crops but low quantities of potassium. In general, soil management does not show any significant problem.

Very well to well-drained, medium fine-textured Fluvents and Fluventic Inceptisols of Karditsa and Trikala areas [(K–T) (Ef + If) (A, B)2]

Deep soils, of medium fine-textured, very well to well-drained, calcareous, with moderate angular or subangular blocky structure, slow infiltration rate (0.1–0.5 cm h^{-1}), with bulk density around 1.33 g cm^{-3} in the surface horizons and good workability and accessibility. Soil pH is neutral (around 7.0) and organic matter content in average is 1.50%. Soil fertility is medium high with concentration of available P (P_{Olsen}) around 16 mg kg^{-1} soil, and available potassium varying around 1.33 cmol(+) kg^{-1}.

This management group is suitable for all the crops grown in the area without serious limitations for any of them. Spring crops require irrigation. Due to favorable pH, all types of fertilizers are suitable.

Well-drained, fine-textured Fluvents of Trikala area (TEfB3)

Deep soils, of fine texture, well drained, calcareous, with moderate angular blocky structure, medium slow infiltration rate (0.5–0.2 cm h^{-1}) and relatively good workability and accessibility. Soil pH is slightly alkaline (around 7.8) and organic matter content in average is 1.92%. This management group is suitable for almost all the crops grown in the area with limitations related to their cultivation due to their fine texture.

Very well to well-drained, coarse-textured Fluvents of Trikala area [TEf (A, B)4]

Deep soils, of coarse texture, very well to well drained, without free calcium carbonate, with weak subangular blocky structure, medium infiltration rate (2.0–6.5 cm h^{-1}), very good workability and accessibility. Soil pH is slightly acid (around 6.4) and low organic matter content in average is 0.89%. This management group is suitable for almost all the crops grown in the area with no serious limitations.

Moderately drained, slightly fine-textured Fluvents + Fluventic Inceptisols of Karditsa and Trikala areas [(K–T)(Ef + If)C2]

Deep soils, of medium fine texture, moderately drained with calcium carbonate in the whole profile or below the surface layers with moderate granular structure in the surface horizon and weak up to moderate subangular blocky structure in the deeper layers with medium low infiltration rate (0.5–2.0 cm h^{-1}). Soil workability as well as accessibility is good. Soil pH is slightly acid to neutral (6.0–7.6), and organic matter content in average is 2.05%. Cation exchange capacity is quite high (in average around 30 cmol(+) kg^{-1}. These soils are suitable for all the crops adapted in the area. Winter crops require drainage to remove surface water.

Moderately to imperfectly drained, fine-textured Fluvents of Trikala area [TEf(C, D)3]

Deep soils, of fine texture, moderately to imperfectly drained with calcium carbonate in the whole profile with moderate crumbly structure in the surface horizon and weak up to moderate angular blocky structure in the deeper layers. Infiltration rate is slow (0.1–0.5 cm h^{-1}). Soil workability is not so favorable due to the fine texture as well as accessibility is not good during the period when surface waters are present requiring drainage by appropriate ways. Soil pH is slightly alkaline (up to 8.0) and organic matter content in average is 2.23%. Cation exchange capacity is quite high (in average around 35 cmol(+) kg^{-1}). These soils are suitable for all the crops adapted in the area but after appropriate improvement of their drainage Winter crops require drainage to remove surface water.

9.6.2 Management Groups of Pieria Region (ISMC 1995)

Pieria is an area located in the northeast part of the country covering in total about 164,000 ha from which 55,000 ha are used in agriculture. The climate is characterized as wet, mesowarm, Mediterranean with the following basic characteristics: mean annual temperature 15.2 °C, mean temperature of the warmest month 25.7 °C, and mean annual temperature of the coldest month 4.8 °C. The mean annual rainfall is 650 mm from which about 560 mm are distributed almost equally in the three seasons fall, winter, and spring and the rest in summer. The most important agricultural crops in the area are the tree crops peach, pear, hazel, and almond, the horticultural crops are strawberries and kiwi, and the grain crops are cereals and corn. In the hilly areas, grapes are cultivated.

Very well to well-drained, moderately fine-textured Fluvents + Fluventic Inceptisols of Pieria area [P(Ef + If)(A, B)2]
This Management groups group covers about 20% of the total agricultural area of Pieria. Deep soils, calcareous in the whole profile or in the subsurface layers of medium texture, very well to well drained. The upper horizons in the cases of Fluventic Inceptisols do not have free calcium carbonate which is concentrated in deeper parts of the soil profile forming in some cases calcic horizons. Soil pH ranges from slightly acid (6.50) up to slightly alkaline (7.90). Organic matter content is quite high i.e., in average 2.11%. Cation exchange capacity is also quite high varying around 35 cmol(+) kg^{-1}. Infiltration rate is slow to moderately slow (0.1–6.5 cm h^{-1}) and bulk density ranging between 1.45 and 1.65 g cm^{-1} in the surface horizon and 1.34–1.45 g cm^{-1} in the deeper layers. Chemical soil fertility is low with exchangeable potassium around 0.20 cmol(+) kg^{-1} and available phosphorus around 4.00 mg kg^{-1}. Workability and accessibility of these soils are very good.

The most important limitation of the tree crops grown in the area (peach, pear, and kiwi) is calcium carbonate content which may cause a reduction of metals uptake, especially iron and manganese causing the respective deficiencies. This limitation is especially serious for the sensitive to calcium tree crops such as peach and kiwi in the soils with calcic horizons and in soils with calcium carbonate in the surface layers for strawberries. Acidifying nitrogen fertilizers are suggested for the tree and horticultural crops and adequate phosphorus and potassium quantities needed. Irrigation of all the crops except cereals is needed which may be applied with any irrigation system.

Very well to well-drained, moderately fine-textured Fluvents + Fluventic Inceptisols of Pieria [P(Ef + If)(A, B)3]
This soil management is similar to the previous one except the soil texture which is heavier resulting in less workable soils. Infiltration rate is slower than in the former management group. Everything referred to the previous management group is applied also to this group except those related to their cultivation which must be done in the appropriate soil moisture content.

Very well to well-drained, coarse-textured Fluvents of Pieria [P(Ef)(A, B)4]
Deep soils, very well to well drained, coarse textured, with carbonates present in the whole profile, or only in the subsurface layers, slightly acid to slightly alkaline pH, and organic matter content ranging from 0.60 to 1.70%. Bulk density ranges around 1.50 g cm^{-3} in the surface horizons and 1.45 g cm^{-3} in the subsurface layers. Workability and accessibility are very good. Soils with low fertility require irrigation and fertilization. Suitable for early crops like strawberries.

Moderately to imperfectly drained, moderately fine-textured Fluvents of Pieria [PEf(C, D)2]
These soils are in the lower parts of the area especially close to the sea where drainage conditions are bad. They are deep soils, calcareous, of medium fine-textured, slightly alkaline to alkaline, and moderately to imperfectly drained. Their workability is good but their accessibility is problematic especially in the periods when there is water on the surface of the soils. In order to grow any crop, it is needed their drainage to be improved. Organic matter content is quite high varying from 1.70 to 4.40%. Quite high is also available phosphorus (between 25 and 40 mg kg^{-1}). After improving their drainage, these soils may become very productive, especially for annual crops. These soils are not suitable for crops very sensitive to drainage conditions such as kiwifruit.

Management groups of Xanthi area (Agricultural University of Athens, 1993)
This area is located in northeast part of Greece and is characterized by a climate with mean annual temperature 15 °C, mean maximum and minimum temperature 29.80 and 1.90 °C, respectively. The mean annual is 725 mm with most rainy month November (118 mm) and most dry month August (32.6 mm). The main agricultural crops are cereals, cotton, tobacco, alfalfa, sugar beets, and industrial tomato and in small areas vegetables and olive trees.

Very well to well-drained, medium-textured Fluvents of Xanthi area [XEf(A, B)1]
Deep soils of medium texture, very well to well-drained, with low free calcium carbonate in the whole depth of the soil profile. The structure of these soils is weak subangular blocky in the upper layers becoming weaker in the deeper layers. Soil pH is neutral up to slightly alkaline (pH 6.5–7.9). Organic matter content is medium ranging from 1.1 to 3.0% in the surface horizons. Cation exchange capacity is medium varying between 15 to 22 cmol(+) kg^{-1}.

Fertility of the soils of this management group is low requiring application of some essential elements. They contain adequate quantities of the essential nutrients Ca but are low in K, as well as the trace elements B and Zn. Available P concentration is adequate.

Soil workability is very good as well as accessibility. For the crops grown in this area, these soils are considered as productive under the appropriate fertilization with the nutrients which are in lower concentrations needed (N, K, B, and Zn).

Moderately drained, moderately fine-textured Inceptisols of Xanthi [XI(C)2]
Deep soils of medium fine-texture, moderately drained, without free calcium carbonate in the surface horizons and

calcareous in the subsurface horizons, moderate to strong medium subangular blocky structure in the whole profile. Soil pH is slightly acidic in the surface horizons becoming higher in the deeper horizons up to 7.9. Organic matter content is low varying between less than 1% up to 1.54% in the surface horizons, and cation exchange capacity is quite high ranging from about 20 to 40 cmol(+) kg^{-1}.

Fertility of the soils of this management group is low requiring application of some essential elements. They contain adequate quantities of the essential nutrients Ca but are low in K as well as the trace elements B and Zn. Available P concentration in most of the areas is adequate, sometimes excessive (>50 mg kg^{-1}). In contrast, K is so low that K fertilization is necessary for all the crops grown in the area. Trace elements both metallic and B are in adequate concentrations. Fertility of these soils is in general higher than the group management of the soil order Entisol.

Soil workability and accessibility is difficult especially in the periods when surface waters exist. Appropriate drainage canals must be constructed. With the appropriate fertilization and irrigation, the soils of this management group is suitable for all the crops cultivated in the area. Potassium fertilization is necessary for almost all the crops of the area and in some cases Zn application is needed.

Very well-drained, coarse-textured Fluvents of Xanthi area [XEf(A)4]

Deep soils of coarse texture (sand loamy) without calcium carbonate in the whole depth of soil profile, with very good drainage. The structure of these soils is weak subangular blocky in the upper layers and weaker up to structurless in the deeper layers. Soil pH ranges from extremely acid (pH 4.5) up to neutral (pH 7.3). Organic matter content is low from 0.8 to 2.0% in the surface horizons and quite lower in the deeper layers. Cation exchange capacity is also low, usually less than 13 cmol(+) kg^{-1}.

Fertility of the soils of this management group is low. They contain small quantities of the essential nutrients Ca and K as well as the metallic nutrients Zn, and Cu but adequate Fe and Mn. Available P concentration is usually adequate. Available B is low in general (0.3–1 mg kg^{-1}).

Soil workability is very good as well as accessibility. For the crops grown in this area, these soils are considered as productive under the appropriate fertilization with the nutrients which are in lower concentrations needed (N, K, B, and Zn).

9.6.2.1 Management groups of Argolida area (Ministry of Agriculture, Agricultural Univesrity of Athens, 1983)

The management groups of this area are located in northeast part of Peloponnese into the prefecture of Argolida. The climate in this area is characterized as warm temperate rainy with the dry period during the summer and warm summer. Mean annual temperature is 17.9 °C and mean temperature of warmest month (July) is 28 °C, while the mean cold temperature of the coldest month (January) is 9.7 °C. The absolute minimum temperature is −1.1 °C (January) and the absolute maximum temperature is 37.8 °C (August). The average annual rainfall is 480 mm being distributed as 40% in winter months while during the summer months it is zero. The main crop in the area is citrus trees and in the second degree vegetables such as tomatoes and eggplants and tree crops as olive trees and apricots.

Very well-drained, medium fine-textured Entisols of Argolida area [Ar_iEf(A)2]

Deep soils, of medium fine texture (mostly clay-loamy), extremely to well drained. Soil structure is weak to medium subangular or angular blocky, and in some cases massive. The soils contain calcium carbonate in the whole profile and have a cation exchange capacity around 15 cmol(+) kg^{-1}. Exchangeable potassium concentration is medium (about 0.50 cmol(+) kg^{-1}) in the surface horizon reduced gradually in the deeper layers. The same trend is true for magnesium which is 4.13 cmol(+) kg^{-1} in the surface horizon. In general, all the exchangeable basic cations are at adequate level for the crops grown in the area. Any nutritional disorder due to these cations should be attributed in the irregular proportion between them rather than the absolute values of their concentration. Manganese concentration as in all soils in the area is in general low due to calcium carbonate content, and possible manganese deficiencies may be found. Copper concentration is in general at adequate level and no deficiencies are expected. Zinc is in general low, and taking into account, the high calcium carbonate content of Cu deficiencies are expectable. Infiltration rate of the soils is medium (mean value 2.8 cm h^{-1}). Bulk density is about 1.12 g cm^{-3} in the surface layers increased in the deeper layers up to 1.50 g cm^{-3}. The soils of this management group are suitable for all the crops grown in the area if they receive the inputs required such as irrigation for orange trees 300–1000 m^3 water.ha^{-1}, nitrogen and in less amount phosphate and potassium fertilizers. A usual problem for the citrus trees is the compacted layer below the plow layer due the incorrect soil management. Usual are also problems related to the incorrect relation between exchangeable Ca and Mg.

Moderately to imperfectly drained, fine-textured Entisols of Argolida area [Ar_{ii}Ef(C, D)3]

Deep soils moderately to imperfectly drained of fine textured found mostly in the lower parts of the alluvium plain. Soil structure is weak to moderate medium in the surface plow layer and moderate to strong medium to coarse in the deeper layers. Calcium carbonate content is present in the whole profile. Organic matter content is the highest in the soils of the area more than 2.5% in the plow layer reduced up to 1% in the deeper layers. The same trend follows nitrogen content which is closely correlated to organic matter content. Exchange K concentration is medium in the surface horizons (about 0.64 cmol(+) kg^{-1} being reduced gradually in the deeper layers up to 0.38 cmol(+) kg^{-1}. Exchangeable Ca is the main basic cation since the soils are calcareous. Exchangeable Mg is high in the whole profile varying around 6.0 cmol(+) kg^{-1}. High values of Mg and the relatively high values of K are attributed to the action of water and the high clay content or in the secondary enrichment. In general, Mg concentration is in adequate level for citrus trees which however in some cases appear Mg deficiency symptoms probably due the incorrect proportion between Mg and Ca. Concerning the concentration of the metallic essential elements Cu and Zn, it can be said that Cu is at relatively high levels (around 7 mg kg^{-1} in the surface layers and about 6.0 mg kg^{-1} in the deeper layers) quite enough to satisfy the needs of the crops grown in the area but Zn is in general low (about 1.3 mg kg^{-1} in the plow layer and about 0.7 mg kg^{-1}) in the deeper layers and Zn deficiencies are possible especially in citrus crops.

Infiltration rate is medium slow (1.5 cm h^{-1}). Mean bulk density is 1.13 g cm^{-3} in the plow horizon increasing in the deeper layers gradually up to 1.47 cm h^{-1}. Workability and accessibility of these soils are bad due to their texture and the presence of waters in the soil surface in some periods requiring drainage management especially for the sensitive to water crops such as citrus trees. In general, the soils of this group are very productive if the appropriate measures described above are taken.

Imperfectly to poorly drained, fine-textured Entisols of Argolida area [Ar_{iii}Ef(D, F)3]

Deep soils poorly drained of fine texture found in the lower parts of the alluvium plain covering a small part of the area of Argolida (about 800 ha). These soils have gley horizons with permanent water table shallower than 150 cm from the soil surface. These are located close to the sea and show extensive salt influence. These soils have strong structure and good hydraulic conductivity and may become very productive if their drainage will be improved.

Very well to well-drained, fine-textured Fluventic Inceptisols of Argolida area [Ar_{iv}If(A, B)3]

Deep soils very well to well-drained of fine texture. The main difference of this management group compared to others is the presence of compacted layer due to the fine structure and structure maturity which creates some limitations such as in the irrigation of the citrus trees.

10 Land Degradation and Desertification

Yassoglou (1986), reported in the European Communities Berlin Symposium on the Scientific Basis for Soil Protection in the European Community, that soil erosion is the dominant degradation process in southern Europe related to soil productivity deterioration. Desertification in Greece is an important process of land degradation especially in the olive and cereal cultivation zone, pine-forested areas, and shrubby grazing lands. In spite of the adverse physical conditions occurred in the country, desertification proceeds only in areas under unsustainableuse of land resources by man. Therefore, desertification, in sensitive areas of Greece, has been and still is an ongoing process. It is due to the over exploitation of soil, water, and vegetation resources. Desertification is processing slowly showing spatial and temporal discontinuation, and is not easily perceived by the various people communities until the land has lost its main productivity irretrievably. It has taken place for more than three millenniums related to the loss of land productivity and availability of water resources. Desertification has been accelerated in the last four decades due to (a) intensification of agriculture, (b) increasing drought due to climate change, and (c) increasing frequency of forest fires.

Land degradation in Greece has been mentioned by the ancient Greek writers. Soil erosion, a process of land degradation and desertification, has been mentioned for the first time in Ilias of Homer (854 B.C.). Platon (fourth century B.C.) has described the region of Attica in which hilly areas were covered by forest in the previous times but due to deforestation resulted in soil degradation. As a term, land degradation is included in many reports of the Byzantine emperor Theodosius (347–395 A.C.) with several references (gobbet 438) to "**deserti agri**" or to the areas abandoned due to low productivity. During the Ottoman period (1457–1913 A.C.), many Greeks have been displaced in the mountainous areas cultivating the land. Many forests have been destroyed during that period for cultivation of the land and used for grazing. As land was degraded, other areas were deforested for the expansion of crops and used for grazing. These actions have been lasted for about four centuries resulting in land degradation and desertification of extensive areas. Mechanization of agriculture during the middle of twentieth century largely contributed to land degradation and desertification. Intensive cultivation of lowland resulted in land abandonment or abandonment of traditional ways of cultivation such as terracing and covering the soil by vegetation. The overexploitation of aquifers resulted in the intrusion of brackish waters and soil salinization of areas especially along the coastal line of the country.

Desertification is a process in which the productive land is gradually degraded leaving spots of land with low productivity with sparse vegetative cover which are expanded and connected forming extensive unproductive areas. Main responsible processes or causes in Greece are (a) the accelerated rates of soil erosion, (b) the degradation of physical, chemical, and biological properties of natural resources, and (c) the loss of natural vegetation. Soil erosion, responsible for land degradation in sloping lands, causes significant decrease of soil depth, fertility soil water storage capacity, and vegetation growth. The climate change is expected to exaggerate the problem since the lack of water is the main threat of the protection of the Mediterranean environment prevailing in Greece. Desertification, besides the adverse environmental consequences, has significant negative economical and social impacts since by degrading natural resources reduces crop production and farmer's income causing rural migration to other areas with greater possibilities of finding a job.

According to the Greek National Action Plan, areas under high risk of desertification are estimated in 34% of the country including part of Sterea Hellas, the greatest part of Peloponnesus, the mountainous areas of the Ionia islands, the islands of the Aegean sea, the island of Evia, the eastern and southern Crete, parts of Thessaly, Macedonia, and Thrace (Fig. 9.1). Areas with moderate risk of desertification occupy 49% of the country, while the rest 16% is considered as subjected to low or no desertification risk.

N. Yassoglou et al., *The Soils of Greece*, World Soils Book Series,
DOI 10.1007/978-3-319-53334-6_10

10.1 Factors and Processes of Desertification

Desertification in Greece as a natural process depends on several factors and processes which act independently or in combination as it was reported by many authors who carried out research on desertification in Greece (Varelides et al. 2004). These factors are natural/environmental, such as climate, physiography, soils, and vegetation or related to socioeconomic or human actions (Fig. 9.2). The climatic conditions of Greece, characterized by long-period drought or intensive rainfalls in combination with sloping topography of the country, favor soil erosion, vegetation degradation, and desertification Human actions such as mismanagement of the land and forest fires aggravate desertification. In addition, the existing Greek or European policies in the last decades have contributed to intensive cultivation of land, overgrazing of pastures, migration of rural population in urban areas, etc. indirectly affecting desertification.

The main problem of land degradation and desertification in Greece is the loss of soil volume capable of supporting vegetative cover. Lands of high quality are limited due to irregular terrain with steep slopes, soil limitations, high rainfall variation, and long periods of misuse. The particular nature of the land with slopes subjected to extensive deforestation and intensive cultivation since ancient times has led to soil erosion and the formation of skeletal soils. According to the results of the CORINE project (1992), 43% of the land is classified as having a high-potential erosion risk, mainly in the south and west, where soils, terrain, and climate combine to create suitable conditions.

The erosion risk is high in areas cultivated with rainfed cereals. For one or two months after sowing winter cereals, the land remains almost bare, and the erosion risk is high considering that rains of high intensity occur during that period. Several sloping lands of the country were for centuries under grazing especially in winter by transhuman flocks and herds. The rapid increase in population due to immigration in early 1920s resulted in land use change from pastures to cereals. Studies carried out by the Agricultural University of Athens demonstrated that erosion on these areas has proceeded at rates of 0.5–1.7 cm soil per year since the period of mechanization of agriculture. A short analysis of the main factors and land degradation and desertification is undertaken in the following paragraphs.

10.1.1 Climate

The climate of Greece belongs to the Mediterranean type which is characterized by great spatial and temporal seasonal and annual fluctuations. The majority of rains fall in the period October–March, whereas the months June–September are characterized almost by the absence of any precipitation. Rainfalls with high intensities in combination with adverse topographic conditions (steep slopes) cause frequently high surface water runoff rates and loss of fertile topsoil with great fluctuation in flow in rivers and creeks and often catastrophic flooding in the lowlands.

According to the National Meteorological Services, the annual rainfall ranges from 1280 mm in western Greece to 370 mm in the eastern part. The western and northern regions of the country receive higher amounts of rainfall in comparison with eastern and southern regions, where the water requirements for plant growth are greater. Rainfall greatly affects surface water runoff and soil loss (Kosmas et al. 1997). Experimental data, collected along the Mediterranean Europe including Greece in hilly areas covered with shrubby natural vegetation, have shown a critical value of annual rainfall in the range 280–300 mm in which the maximum water runoff was observed. As annual rainfall increased, vegetation cover increased and annual surface water runoff decreased (Kosmas et al. 1997).

Rainfall amount and distribution are the major determinants of biomass production on hilly lands under the climatic conditions of Greece. Decreasing amounts of rainfall combined with high rates of evapotranspiration drastically reduce the soil moisture content available for plant growth. Reduced biomass production, in turn, directly affects the organic matter content and aggregate stability of the surface soil horizon, therefore sensitivity to soil erosion.

Aridity is related to the average water available in the soil. The higher the aridity index of a region, the greater the water resources' variability and scarcity in time, the more vulnerable the area to desertification. Studies carried out in the island of Lesvos (eastern Greece) have shown that aridity has a great influence on vegetation performance.

10.1.2 Soil

Soil is one of the most important factors affecting vegetation growth in the semiarid and dry subhumid climatic conditions of Greece. When soil cannot provide to the plants the necessary amount of nutrients and water, then desertification is expected. The most important soil parameters affecting desertification in Greece are soil depth, parent material, rock fragments, and drainage Desertification becomes irreversible when the soil cannot support the minimum plant cover of 45–50%. Percentage vegetation cover of a soil surface under semiarid climatic conditions is largely controlled by soil water storage capacity and therefore soil depth. Soil depth decrease due to soil erosion is a serious threat to the soil quality and productivity in hilly areas in Greece. Studies carried out in the island of Lesvos have shown that there is a critical depth of 30 cm (Fig. 9.3), depending on the type of parent material, which can support the minimum 45%

vegetation cover for reduction of soil erosion to low rates (Kosmas et al. 2000).

Soils present different sensitivities to soil erosion and desertification depending on the parent material. Shallow soils are mainly formed on limestone parent material favoring high soil water deficits due to the presence of cracks in which rainwater is percolating deeper. In the opposite, soils formed on flysch parent material are usually deeply favoring vegetation growth. Many areas in Greece with soils formed on limestone have been desertified. Similarly, soils formed in acid igneous rocks are shallow with high desertification risk. In the opposite, areas with soils formed on shale parent material are normally characterized by high productivity, supplying appreciable amounts of previously stored water to the stressed plants protecting areas from desertification (Kosmas et al. 1993). On the other hand, soils formed on marl, despite their considerable depth and high productivity in normal and wet years, they are very susceptible to desertification and unable to support any annual vegetation, particularly in dry years due to adverse soil physical characteristics. These soils frequently form a surface crust inhibiting water infiltration favoring surface water runoff.

The presence of rock fragments on the soil surface affects the following significantly: (a) surface water runoff and sediment loss (Danalatos et al. 1995), (b) soil water conservation, (Moustakas et al. 1995), and (c) biomass production, therefore, sensitivity of the land to desertification. Field experiments carried out by the Agricultural University of Athens have shown that rock fragments on the soil surface with dimension greater than 5 cm favor surface water runoff; in the opposite, rock fragments of smaller dimensions in an amount greater than 20% reduce surface water runoff. Furthermore, rock fragments affect positively soil water conservation under moderate conditions of water deficit as those prevailing in Greece during spring and summer. Soil water evaporation is reduced significantly when rock fragments are present on the soil surface. Wheat biomass production increased by 10–30% in hilly areas of Thessaly and Viotia after the removal of rock fragments from the soil surface, compared to the nearby areas without removing stones (Danalatos et al. 1995).

Soil drainage is related to how rapidly excess water leaves the soil by runoff or internal drainage. Soil drainage is especially important in lowlands located especially along the coast in which the intrusion of brackish water is favored by overexploitation of the aquifers for irrigation. Areas in Greece vulnerable to soil salinization and desertification are mainly the Argos plain (Peloponnesus), the Acheloos river plain (Sterea Hellas), Thessaly plain, lower area of Axios river plain (Macedonia), lower Nestos river plain (Macedonia), lower Evros river plain (Thrace), and lower Kalamas river plain (Epirus).

10.1.3 Topography

The greatest part of Greece is mountainous with steep slopes with plain areas formed in valleys after the deposition of sediments by the rivers. Actually, about 49% of the country has slopes greater than 10% and only 36% of it comprises lowlands with slope less than 5%. Areas at elevation greater than 800 m occupy 28.6% of the country. The highest elevation is 2917 m on the Mount Olympus. The main parameters related to topography and affecting land degradation and desertification in Greece are slope gradient and slope exposure.

Slope gradient greatly affects amount of surface water runoff and soil sediment loss. Soil erosion rates become acute when slope angle exceeds a critical value and then increases logarithmically. Studies conducted in the island of Lesvos (Greece) have shown that soil erosion was moderate or high in slope gradients greater than 12%. Slope gradient can have variable effect in different climatic zones, depending mainly on annual rainfall.

Slope aspect is considered an important factor for land degradation affecting the microclimatic conditions by regulating the angle and the duration in which the sun's rays strike the surface of the soil. Variation in slope aspect and elevation influence the distribution of energy, meteoric water, plant nutrients, and vegetation growth by varying the exposure of the soil to wind and precipitation and the conditions for natural drainage and soil erosion. In Greece, slopes with southern- and western-facing aspects are warmer and have higher evaporation rates than northern and eastern aspects. Therefore, a slower vegetation recovery, on southern and western aspects, result to higher erosion rates, than that on the northern and eastern aspects. As a consequence, southern-exposed slopes usually are more degraded or are more sensitive to desertification than northern-exposed slopes.

10.1.4 Vegetation

The main biotic factor (except man) affecting desertification in Greece is vegetation. Vegetation is crucial factor affecting soils in all its dynamics including erosion control, water redistribution over and within the soil, and the microbial activity. Surface water runoff is greatly controlled by vegetation cover which can be readily altered in hilly areas depending on climatic conditions, land management practices, and period of the year. Areas covered in the past by natural vegetation are continuously decreased and replaced mainly by agricultural crops and tourist installations. The total area covered by forest today is estimated to 20% of the territory compared to 48% in 1825. It is estimated that the natural vegetation is decreased in a rate of 0.076% yearly

(Alexandris 1989). The high rates of deforestation in combination with the high intensity of cultivation of the sloping land since ancient period resulted in accelerated soil erosion and land degradation and desertification.

Vegetation and type of land use are important factors affecting intensity and frequency of surface rainwater runoff and sediment loss. Extensive hilly areas in Greece with moderate deep or shallow soils are cultivated with cereals, vines, almonds, and olives. Removal of the understory annual vegetation from vineyards, almond plantations, and olives groves creates favorable conditions for overland flow and soil erosion (Fig. 9.4). Measurements of degree of soil erosion carried out in the island of Lesvos in areas covered with pines, oak trees, and olive groves have shown that pine-forested areas in southern slope exposures have moderately protected soil from erosion due to greater rate of decomposition of dead leaves and lower vegetation growth compared to the northern slope exposures. Oak trees protect inadequately soil especially in areas where the fallen leaves did not cover the soil surface in a great extent. Furthermore, studies carried out in the area of Attica (Kosmas et al. 1995) have shown that the replacement of olive groves existing for more than 160 years by vines resulted in soil aggregate stability deterioration and soil organic matter decrease by 33% in a period of 12 years.

Drought resistance of the growing plants, affecting desertification, is related to the type of vegetation. The majority of the natural growing vegetation and some of the agricultural crops are highly adapted to dry climatic conditions. Mediterranean macchia, pines, and perennial grasses are well adapted to dry climatic conditions. Among the agricultural crops, olive plantations are particularly well adapted to prolonged dry climatic conditions supporting high biodiversity. In addition, the presence of leaves during the whole year adequately protects soil from erosion, land degradation, and desertification.

Both runoff and sediment loss are related to plant cover. Soil erosion decreases exponentially with increasing percentage of plant cover. In addition, plant cover reduces significantly summer soil surface temperatures and soil water conservation. Plant cover is a crucial indicator of land desertification especially in areas affected by water stress. Data collected in the island of Lesvos have shown that plant cover increased with increasing soil depth and decreasing droughts. The class 25–50% plant cover had higher frequency (93%) in semiarid climatic zone in soils with depth 15–30 cm. In the opposite, in the dry subhumid climatic zone of the island under the same soil depth class, the plant cover 75–90% had a frequency of 64%.

10.1.5 Human Actions

The biophysical factors mentioned above constitute significant reason of land degradation of the sensitive ecosystems to desertification. However, the irrational intervention of man leads to irreversible conditions. Human actions directly related to desertification are unsustainable land management practices such as destruction of natural vegetation, overexploitation of water and soil resources. Destruction of natural vegetation and expansion of agricultural areas or pastures in low productivity soils are probably the most important human intervention accelerating the rate of desertification. Land in hilly and mountainous areas has greatly degraded due to overgrazing. The number of animals grazing the land is two or three times higher than the sustainable number leading to the deterioration of soil aggregate stability, alteration of biodiversity, and change in grass production. The presence of long droughts occurring repeatedly in the country leads to reduction of grass production in combination with the unsustainable number of animals grazing the land resulted in irreversible land degradation and desertification.

Forest fires have increased dramatically the last decades. According to the Greek Ministry of Agriculture, the average burned area was 12,900 hectares in the period 1964–1975 and was increased to 37,800 hectares in the period 1976–1986. In the present years, the average of burned areas has increased to 52,358 hectares. The highest rate of burned area occurred in the year 2007 in which 250,000 ha of natural and agricultural land has been burned (Fig. 9.5). The rates of soil erosion and land degradation, in burned areas, are especially high in the following the fire two years period. When the frequency of fire in the same areas is high accompanied by grazing, degradation of soil and vegetation is irreversible.

Degradation of surface and groundwater is obvious in the last decades due to agriculture, touristic and industrial activities. The irrigated land is estimated to 32% of the agricultural land. The intensification of agriculture accompanied by the use of poor quality of water (seawater intrusion) is in many cases responsible for soil degradation resulting from water logging, salinization, and alkalinization. Soil salinization resulting from poor quality of water is one of the key processes that could lead to desertification especially in the plain areas along the coast. It is estimated that 15% of the agricultural land has been affected or will be affected by high concentrations of soluble salts.

The overuse of fertilizers in the previous decades resulted in soil acidification especially in soils formed in acid parent materials. Acidification of soils with pH values lower than 4.5 causes severe toxicities to the growing plant leading to desertification. It is estimated that 450,000 ha of agricultural soils face problems of acidification.

Other human actions affecting indirectly land degradation and desertification are socioeconomically related to land use change in high populated areas such as Attica. Migration of rural people to urban areas has led to land abandonment of hilly areas or to intensive cultivation of plain areas. The Common Agriculture Policy has greatly affected land uses and rate of soil resources' exploitation. The lack of adequate policies for protecting sensitive areas to desertification and to prohibit the irrational and uncontrolled degradation of natural resources has greatly contributed to desertification in a significant part of the country.

10.2 Measures to Combat Desertification

As it has been described above, and water availability are the major issues of desertification in Greece. Measures for soil protection and water conservation should not be limited to a single land management practice, but it requires a number of actions, such as retaining the vegetation cover, applying conservation tillage, hanging the land use, water harvesting, grazing control, forest fire protection, etc. Conservation tillage has been shown to reduce greatly rill and interill erosion in hilly areas cultivated with cereals. The incorporation of plant residues into the soil after harvesting is greatly restricted due to the very dry soil conditions during the summer period. In addition, contour farming is applied to large areas cultivated with winter crops.

Terracing is an alternative land management practice for soil and water conservation and erosion control. Terracing decreases the slope length and reduces the damage by surface runoff (Fig. 10.1). It is actually a very old practice having been used for centuries. Many uplands have been terraced for cultivating cereals, vines, olives, and other crops throughout the country. In many cases, terraces which have been constructed with stones are hundreds or even thousands years old.

An effective measure that might combat further land degradation is the development of alternative environmentally sound and socially tolerable land use schemes. These schemes should be implementing zoning in each threatened area (Yassoglou and Kosmas 1988). Slope gradient and elevation should be the main criteria for the delineation of each management zone (Figs. 10.2, 10.3, 10.4, 10.5 and 10.6).

Water harvesting is mainly related to the constructions of water reservoirs and irrigation systems, enrichment of

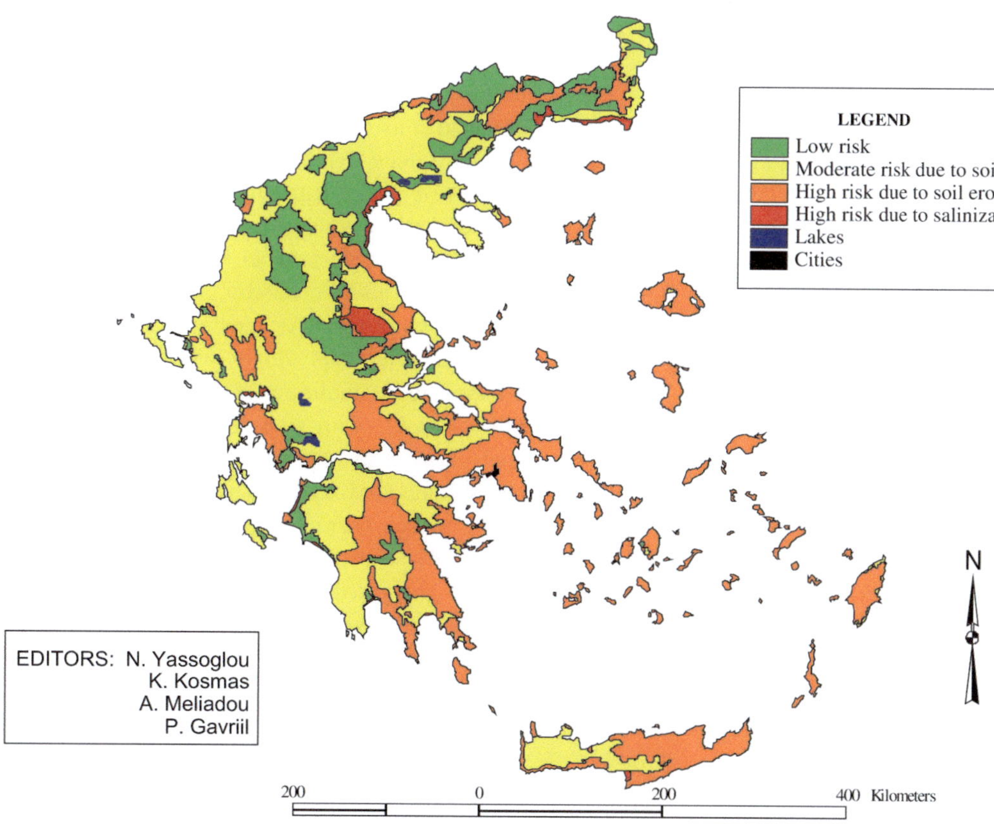

Fig. 10.1 Potential desertification risk map of Greece (Greek National Committee to Combat Desertification)

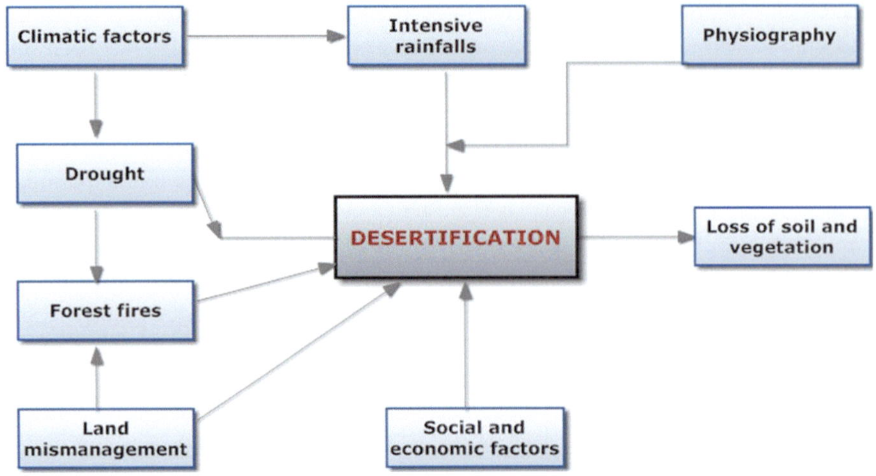

Fig. 10.2 Schematic presentation of the main factors affecting desertification in Greece

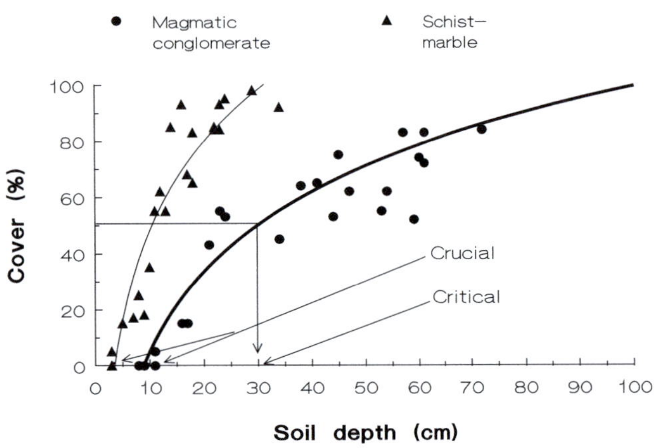

Fig. 10.3 Relation of percentage plant cover provided by the plant *Sarcopoterium* sp and soil depth measured in various parent materials (Kosmas et al. 2000)

aquifers, enhancing water vapor adsorption, storing water in ponds, etc. In addition, understory vegetation in olive groves is allowed to grow during the winter period highly restricting rainwater runoff. Considering that the amount of rain occurring under semiarid climatic conditions is about one half compared to the evapotranspiration rates during the dry period (May–October), water vapor adsorption becomes one of the most important sources of available water for the growing vegetation. Under such climatic conditions, soil physical characteristics such as surface mulching, density of the growing plants, and plowing the soil for increasing macroporosity greatly affect water vapor adsorption and soil water conservation.

Grazing control is related to actions such as sustainable number of grazing animals and fencing of grazing land taken in a pasture for land protection. However, the existing policies on subsidies do not favor such actions. In addition, grazing control is negatively affected by the high grazing land fragmentation.

Generally speaking, Greek agriculture is affected by the European Union Common Market policy. According to the Regulation 797/85, member states are explicitly permitted to introduce their own national aid schemes for supporting farming practices that preserve or improve the environment. The Regulation 1760/87 of the European Commission encouraged the member states to define the areas of sympathetic agricultural practices. The Maastricht Treaty (1992) recognized that the European Union must promote measures at an international level to deal with environmental problems and ensure "sustainable growth respecting the environment." It has been planned that 15% of the land cultivated with crops such as cereals and sugar beets is to be put in fallow.

The Greek Ministry of Agriculture has defined the "Codes of Good Agricultural Practice" in which practices are defined for environmental and human health protection. These codes interfere in a very wide spectrum of the agricultural and husbandry activities as well as to other specialized cases. They include the following: plant coverage in soils with slope over 10%, strip rotation along contour lines, minimum plowing or plowing along contour lines in fields with slope over 10%, alternating cultivation and fallow strips, use of non-soil eroding irrigation systems on sloping land, reduction of grazing density, and prohibition of the burning of crop residues in soils with slope over 10%. Subsidies to those exercising biological agriculture are given on the condition they apply the above measures.

The integrated land management applied mainly to olive groves is a type of land management which secures sustainability of natural resources combined with maximum crop production, promote biodiversity, and protect farmer's health. The application of this response is controlled by the

Fig. 10.4 Agricultural areas intensively cultivated with almond trees (*up*) and vines (*down*) in which understory vegetation has been removed favoring soil erosion

local authorities and farmer's received extra payment. The integrated land management practice has been initiated in the last decade.

The UN Convention for Combating Desertification was ratified by the Greek Parliament on February 28, 1997. Under the ratified convention, the National Action Plan for combating desertification in Greece has been developed which provides main guidelines and mechanisms for combating desertification at national level. The plan in its final form has incorporated the suggestions and improvement proposed by the stakeholders, public authorities, and non-governmental organizations, in the text of the Provisional Action Plan, which was sent to them. The Greek Government accepted officially the National Action Plan in July 2001, through a Common Ministerial Decision (CMD) of six involved ministers.

10.3 Research Related to Desertification

Research related directly or indirectly to desertification has been carried out by the various Greek institutes and universities by funds provided mainly by the following: (a) the ministries of Rural Development and Food,

Fig. 10.5 Satellite image of summer 2007 (*above*) showing extensive forest fires and burned area subjected to severe erosion after this fire (*down*)

Environment, Public Works and (b) European Union. The main research projects on land degradation and desertification were related to the following topics:

- Assessment of factor and processes of Mediterranean desertification and deriving methodologies for identifying and mapping sensitivity to desertification (EU projects MEDALUS I, II, and III). A methodology has been derived for assessing environmentally sensitive areas (ESA methodology) to desertification.
- Development of a system of desertification indicators or Mediterranean Europe (DESERTLINKS). A decision support system (DIS4ME) has been derived to (a) identify where desertification is a problem, (b) assess how critical the problem is, and (c) understand better the processes of desertification.
- uture of olive plantation systems on sloping and mountainous land and management scenarios for production and natural resources' conservation (OLIVERO) and development of decision support systems on integrated management of olive groves and sustainable use of water resources (IMAGE).
- Definition of hot spots in Greece based on the available soil, vegetation, climate, topography, and management

Fig. 10.6 Land terracing and protection of understory vegetation in olive groves for and land desertification protection

characteristics using the Pan-European Risk Assessment model PESERA.

- Desertification, Mitigation, and Remediation of Land—a global approach for local solutions (DESIRE). A methodology was derived using indicators for assessing land management practices and select the appropriate technology for combating desertification.
- Land and ecosystem degradation and desertification—assessing the fit of responses (LEDDRA). Responses to land degradation and desertification were assessed based on resilience of the ecosystem, capital characteristics, and lower level properties (robustness, redundancy, diversity, etc.).
- Development of surveillance system for assessing and monitoring desertification (DESURVEY, GEOLAND) using especially remote sensing data.
- Assessment of policies and land uses for combating desertification (MEDACTION). The various existing policies have been analyzed and recommendations for improvement have been made for combating desertification.
- Support to regional action plans of northern Europe for combating desertification (MEDRAP). An effort has been undertaken to combine national action plans to desertification of Annex IV countries into the regional action plan for southern Europe.

- Soil quality assessment for assessing agricultural productivity and environmental resilience using indicators (iSQAPER). Indicators have been used for assessing land management practices and degradation processes for sustainable use of soil resources.

Finally, the Ministry of Education is continuously upgrading the educational program in all its levels with particular emphasis on the construction of new schools, filling personnel gaps, and improving tertiary education and enhancing environmental issues.

11 Soil Research

Greek people had been observing the properties and the functions and the benefits of soil Gaia (Γη) since the classical times. They had developed appropriate systems and practices for sustainable land use based on the evaluation of land capability.

Greeks like the Juish people had regarded the soil and its use in a religio-philosophical frame (Yassoglou 2015).

The rigorous scientific study of the soil was initiated and conducted during the first decades of the twentieth century by a group of scientists led by the Russian immigrant Zvorykin in the frame of the Kanelopoulos Chemistry Institutes established by the homonymous fertilizers industries.

Generally speaking, education in agriculture has been initiated with the establishment of the Greek Nation after deliberation from the Ottomans (1821). The first education agricultural school has been founded in 1829 in Tirintha (Peloponnesus). Education and research on soils has been basically initiated with the establishment of the first university of Higher Agricultural School of Athens (1920) and Agricultural School of Aristotle University of Thessaloniki (1927). The Soil Institutes of Thessaloniki and Athens have been established in 1950 and later (1981) the third Institute of Soil Mapping and Classification of Larissa. In addition, a soil laboratory has been established (1961) in the Center of Nuclear Research Democritus in Athens. In the last two decades, more soil laboratories have been established at university level such as the laboratories of Soils in the Universities of Thessaly and Thrace. Research has been carried out in various topics related to soils such as soil mapping and land evaluation, soil genesis and classification, soil use and land management, soil degradation and land protection, soil fertility and fertilization, chemical and physical properties, soil quality and groundwater pollution, and application of wastes in soils.

11.1 Forest Soils

Extensive research projects on soil properties, conservation, and suitability of forest soils were conducted by the School of Forestry of the Aristotelian University of the Thessaloniki and the Forest Research Institutes of Athens and Thessaloniki of the Ministry of Agriculture. Representative results and scientific data obtained by these efforts are included in the related chapters of this book (some relevant references are as follows: Alexandris 1989; Nakos 1979, 1983, 1984, 1987; Papamichos 1979; Alifragis and Papamichos 1992; Gakis et al. 2014; Mpokos et al. 2014).

11.2 Soil Mapping and Land Evaluation

A semi-detailed (scale 1:75,000) soil survey has been initiated in Thessaly by the soil scientist Zvorykin in 1931 using the Nabochik soil survey system. About three decades later, a detailed (scale 1:10,000) soil survey has been carried out by the Yassoglou in the selected areas of Greece (Messara Valley, Argos Plain, Pinios River Plain, Evrotas River plain). The national project on soil mapping has been initiated in 1977 for a detailed (1:20,000) soil mapping of agricultural soils and semi-detailed (1:50,000) mapping of forestry soils by the Ministry of Agriculture in collaboration with the Agricultural University of Athens. This project has been technically and economically supported by the Food and Agricultural Organization (FAO) of the United Nations and European Commission. The American soil classification systems were adopted for mapping the soils of Greece. A parametric system of soil mapping has been established that applied in all soil surveys carried out after 1977. All these efforts resulted in compilation of soil survey reports of selected agricultural areas such as Thessaly, while the soil

N. Yassoglou et al., *The Soils of Greece*, World Soils Book Series,
DOI 10.1007/978-3-319-53334-6_11

survey of forestry soils has been finalized. Finally, another project has been funded by the European Union in 2013 for creating a soil database for the 3,850,000 ha in the scale 1:30,000.

The Ministry of Agriculture carried out extensive soil studies mainly to support land reclamation projects and/or irrigation and drainage projects executed in the main agricultural areas such as Thessaly Plain, Macedonia Region, Thrace, Epirus, Sterea Hellas, Peloponnesus, and Crete. These studies included the main soil properties that affect the flow of water in the soil such as texture, limiting soil layers, salinization, carbonates, and drainage. Such studies have been carried out in several regions of the country initially by the Ministry of Agriculture and later by the established soil institutes. Soil survey reports have been compiled, and numerous publications have been published in international journals and conferences.

Land evaluation for agricultural crops has been investigated by using detailed models, pedotransfer functions, and parametric systems by various researchers. The derived empirical systems have been used for assessing suitability of soils for various crops growing under the climatic conditions of Greece. Recently, a database has been derived for the majority of agricultural soils in which land users can identified the suitability of their farm for a large number of crops.

11.3 Soil Genesis and Classification

The formation of soils in various parent materials has been investigated. Red soils, a widely expanded category of soils in the country, have been investigated concerning soil chemical characteristics, parent material, drainage conditions, and climatic conditions (Yassoglou et al. 1997). Soil formation has been investigated using archeology findings, and the minimum time period for the formation of cambic and argillic horizon under the semiarid climatic conditions was determined. The soil and climatic conditions prevailing in the country have been investigated in relation to iron and aluminum oxides, and hydroxides (Haidouti and Yassoglou 1982; Sgouras et al. 2007). The process of cracking of soils containing high amounts of expanding clays has been investigated in research projects in relation to water bypass flow. The parent material, the topographic conditions, and the weathering process on the formation of cracking soils classified as vertisols have been studied in various regions of the country (Moustakas 2012). Genesis and classification of soils formed on volcanic parent materials have been extensively studied by various researchers (Moustakas and Georgoulias 2005; Drouza et al. 2007; Georgoulias et al. 2010; Mizota et al. 1988).

11.4 Soil Degradation and Land Protection

The main threats of soil degradation in Greece such as soil erosion, soil salinization, organic matter loss, heavy metal contamination, and soil compaction have been investigated by various researchers. Long- or medium-term soil erosion experiments due to surface water runoff have been installed in various places of the country, and the main factors, such as land use, plant cover, parent material, presence of rock fragments, topography affecting water runoff, and sediment loss have been investigated. In addition, land management practices have been assessed in relation to sustainable use of soils. Furthermore, soil erosion due to the displacement caused by the cultivation instruments such as harrow disk, moldboard, and cultivator has been studied in relation to slope gradient, slope shape, tilling depth, and direction, and pedotransfer functions have been derived (Kosmas et al. 1993, 2000).

Soil salinization in relation to land characteristics and management practices and exploitation of aquifers has been investigated in various plain areas of Greece. Leaching processes and water requirement for leaching have been studied in relation to soil texture, porosity, hydraulic conductivity, depth to groundwater, etc. Soil salinization and electrical conductivity are monitored in several alluvial plains, and projects of groundwater enrichment have been funded (some publications on this issue are as follows: Misopolinos et al. 1988; Misopolinos 1985, 1987).

Organic matter loss under various land use, land management practices, soil characteristics, and climatic conditions has been widely studied. Sustainable land management practices or land uses have been investigated and proposed to the farmers for soil organic carbon conservation. In addition, the processes of organic matter decomposition, important organic matter substances, rate of decomposition, and contribution to soil aggregation have been investigated (some relevant references are as follows: Moustakas et al. 1995; Marathianou et al. 2000; Kairis et al. 2013).

Heavy metal contamination of soils such as lead, nickel, cadmium, zinc, and arsenic has been extensively investigated. The various sources of heavy metal contamination of soils, such as gas, fertilizers, pesticides, and herbicides as well as domestic or industrial waste disposals have been studied. In addition, the mechanisms of heavy metal adsorption in relation to various soil characteristics have been analyzed (some relevant papers on this subject are as follows: Kanellopoulos et al. 2015; Tsadilas et al. 1995, 2014; Kelepertzis 2014; Tsadilas and Shaheen 2010, 2013; Antoniadis et al. 2010; Skodras et al. 2007; Matsi and Keramidas 1999; Shaheen and Tsadilas 2013a, b).

Soil compaction of the subsurface horizon and soil crusting of surface horizon are major problems in agricultural soils. Several studies on aggregate stability, soil porosity, and bulk density have been carried out during the execution of soil surveys. Relations between amount of organic matter, land use type, and land practices have been formulated, and measures for decreasing compaction have been proposed. Water vapor adsorption has been investigated and related to soil physical and climate characteristics (examples of relevant research are the papers: Panagiotopoulos et al. 1994; Lithourgidis et al. 2009).

11.5 Soil Fertility and Fertilization

Soil fertility and fertilizers of the soils of Greece have been extensively investigated. A significant number of studies have been oriented to soil amelioration through the application of soil amendments as well as nutrients uptake by plants grown on the ameliorated soils. Nutrients under consideration included phosphorus, potassium, nitrogen, boron, and metallic micronutrients. Phosphorus is among the nutrients drawn the greatest attention of the researchers in the country since it is one of the basic essential nutrients for plant growth especially those soils classified as Rhodoxeralfs, Palexeralfs, Xerochrepts, and Orthents due to high P adsorption capacity causing very often phosphorous deficiency in the growing crops. The one- and two-surface Langmuir and Freundlich isotherms of phosphorous sorption have been investigated by many researchers. In addition, the various forms of phosphorous in forest soils of Greece have been studied and related to parent material. The adsorption of phosphorous by various soil constitutes such as iron oxides, carbonates, organic matter, and clay content has been studied. Various methods of phosphorous extraction have been investigated and related to the availability of the growing plants (some respective publications are as follows: Polyzopoulos et al. 1985; Nakos 1983; Tsadilas et al. 1996; Shaheen et al. 2009; Noulas et al. 2013).

Potassium relationships between quantity and intensity have been investigated in relation to the presence of other ions such as calcium and ammonium. The application of zeolite, a mineral rich on potassium, has been studied in various soils, and the effectiveness on plant growth has been assessed. In addition, studies have been carried out on the effect of soil moisture on the adsorption of potassium by expanding clay minerals (some relevant publications: Barbayiannis et al. 1996; Tsadilas and Evangelou 1998; Assimakoloulos et al. 1994).

Extensive studies have been carried out on the nitrogen flow and availability to the growing plants. The various processes such as nitrogen mineralization, ammonium volatilization, and nitrogen denitrification have been studied under various soil, land use, and climatic conditions. Particular efforts have been made on the assessment of the various soil characteristics and land management practices on nitrate leaching and water pollution. Several projects have been funded on nitrogen fertilization and plant growth (as examples see the papers Kosmas et al. 1991; Yassoglou et al. 1994; Karyotis 1994).

Boron as an essential micronutrient has drawn the attention of many researchers due to the problems that cause to the plants both deficiency to high boron-demanding crops and toxicity to boron-sensitive plants. The various soil boron fractions have been investigated in relation to the plant availability. In addition, the various forms of soil boron have been correlated to soil properties such as organic matter, acidity, iron, and aluminum hydroxides. Furthermore, many researches have been developed or assessed existing methods of boron determination in relation to plant availability (some relevant publications are as follows: Tsadilas et al. 1994, 1997; Tsadilas and Chartzoulakis 1999b; Tsadilas and Kazai 2005; Dionisiou et al. 2006.).

11.6 Soil Chemical and Physical Properties

The geochemistry of various parent materials has been analyzed and compared to the concentration of heavy metals such as cobalt, chromium, nickel, and lead. Arsenic has been measured in soils and especially in waters for defining the quality of water. Furthermore, heavy metal pollution has been investigated in urban areas and compared to soil and traffic characteristics. In addition, vanadium has been studied in various soils for defining the level of contamination. Lead contamination has been investigated in lead mining areas of the country (some relevant references are as follows: Kanellopoulos et al. 2015; Kelepertzis et al. 2006, 2014; Tsadilas et al. 2014; Tsadilas and Shaheen, 2010; Papadopoulos et al. 2007, Tsatiris et al. 2008).

Physical properties such as soil moisture characteristics have been studied, and empirical relations have been derived for assessing the soil moisture retention curves based on the easily available soil properties such as clay and silt content, organic matter, and soil classification. Cracking characteristics of clayey soils and by-pass flow of fertilizers have been studied in various soils. Bulk density, porosity, compaction, and hydraulic conductivity have been measured in various soils for defining land management characteristics, especially irrigation and drainage systems. Soil salinization in relation to chemical and physical properties has been investigated. Critical values of salt content have been identified in relation to the tolerance of the growing plants (Kosmas et al. 1991; Yassoglou et al. 1994).

11.7 Soil Quality and Groundwater Pollution

Soil quality indicators have been studied and related to soil threats such as soil erosion, land desertification, and groundwater pollution or soil functions such as biomass production. Basic land parameters such as soil texture, soil depth, topography, parent material, and climatic conditions were related to the vegetation performance and abandonment of agricultural land under various climatic conditions. The presence of rock fragments has been investigated in relation to soil water conservation and biomass production. Changes in chemical properties have been studied in relation to various land uses (some references on this subject are as follows: Tsadila et al. 2012; Evangelou et al. 2014; Kosmas et al. 1996, 1998, 2000a, b; Zalidis et al. 2002).

Nitrogen leaching and groundwater pollution have been extensively investigated in relation to land management, soil, land use, and climatic characteristics. Based on the studies carried out on nitrate leaching, the National Action Plan for the protection of groundwater and surface water from nitrate pollution has been compiled. Furthermore, the mechanisms of phosphorous adsorption and movement into the soils have been investigated in relation to eutrophication of surface and groundwaters.

11.8 Waste Disposal in Soils

A serious source of point pollution of soils is the disposal of industrial and municipal wastes. Many researchers in Greece have carried out extensive research during the last decades on the impacts on physical and chemical properties and plant growth characteristics. The effect of waste application on soil organic matter content, soil acidity, electrical conductivity, plant fruit production, and quality of production has been investigated in a variety of soil and land use characteristics. The environmental impacts of wastewater application on soils have been investigated. In addition, the application in soils of the industrial by-product ash produced by burning the coal for electricity production has been investigated in relation to chemical and physical properties. Some research has been also conducted on the remediation of polluted soils by using appropriate plants (phytoremediation) (some relevant publications are as follows: Samaras et al. 2008, 2010; Koukouzas et al. 2006; Tsadilas et al. 1999, 2002, 2009; Shaheen et al. 2010, 2014; Suchkova et al. 2010).

Appendix

LEfA1: (Larissa, Fluvents, very well drained, medium textured)
LEf(A, B)2: (Larissa, Fluvents, very well to well drained, medium fine textured)
L(Ef + If) (A, B)3: (Larissa, Fluvents, very well to well drained, fine textured)
LEfA4: (Larissa, Fluvents, very well drained, coarse textured)
LEfC2: (Larissa, Fluvents, moderately drained, medium fine textured)
LEfC3: (Larissa, Fluvents, moderately drained, fine textured)
(K − T) (Ef + If) (A, B)1: (Karditsa-TRikala, Fluvents, Fluventic Inceptisol, very well to well drained, medium textured)
(K − T) (Ef + If) (A, B)2: (Karditsa-TRikala, very well to well drained, medium fine textured)
TEf B3: (TRikala, well drained, fine textured)
TEf (A, B)4: (TRikala, very well to well drained, coarse textured)
(K − T) (Ef + If)C2: (Karditsa, TRikala, Fluvents+ Fluventic Inceptisols, moderately drained, slightly fine textured)
TEf(C, D)3: (TRikala, moderately to imperfectly drained, fine textured)
P(Ef + If) (A, B)2: (Pieria, Fluvents + Fluventic Inceptisols, very well to well drained, moderately fine textured)
(Ef + If) (A, B)3: (Pieria, Fluvents + Fluventic Inceptisols, very well to well drained, moderately fine textured)
P(Ef) (A, B)4: (Pieria, Fluvents, very well to well drained, coarse textured)
PEf(C, D)2: (Pieria, Fluvents, moderately to imperfectly drained, moderately fine textured)
XEf(A, B)1: (Xanthi, Fluvents, very well to well drained, medium textured)
XI(C)2: (Xanthi, Inceptisols, moderately drained, moderately fine textured)
XEf(A)4: (Xanthi, Fluvents, very well drained, coarse textured)
Ar_iEf(A)2: (Argolida Entisols, very well drained, medium fine textured)
Ar_{ii}Ef(C, D)3: (Argolida Entisol moderately to imperfectly drained, fine textured)
Ar_{iii}Ef(D, F)3: (Argolida Entisols imperfectly to poorly drained, fine textured)
Ar_{iv}If(A, B)3: (Argolida Fluventic Inceptisols very well to well drained, fine textured)

N. Yassoglou et al., *The Soils of Greece*, World Soils Book Series,
DOI 10.1007/978-3-319-53334-6

References

Agricultural University of Athens. 1993. Soil Survey Study of the Agricultural Soils of the Pefecture of Xanthi. Pp. 145.

Alexandris, S., 1989. The impacts of fires on the forestry and natural environment. In: The protection of the environment and the agricultural production. Geotechnical Society of Greece, Thessaloniki, 353–364.

Alifragis, D.A. and N.T. Papamichos. 1992. Nutrient dynamic and organic-matter production in an oak forest ecosystem in North Greece. Proc. Of the 1st European Symposium on terrestrial ecosystems: Forests and wood land, Florence, Italy May 20–24, 1991. Pp. 666–668.

Amemiya, T. 2007. Economy and Economics of Ancient Greece. Routledge Explorations in Economics History. pp. 209.

Anestis, G. and Perris, I. 1986. Ministry of Agricultural Development and Foods, Greek. Forest Service Soil Data Base.

Antonakopoulos, G. 1978. Translocation of clay and Fe-Mn oxides in imperfectly drained soils on Pleistocene deposits. Doctors Thesis. Agricultural University of Athens {Grk}, Athens, Greece. 152pp.

Antonakopulos, G. and Yassoglou, N. G. 1986. Distribution patterns of clay and "free oxides in polygenetic soil profiles on Pleistocene Surfaces in Greece. Transactions XIII Congress of the international society of soil science. Extended informative summaries. 1034–1035.

Antoniadis, V., C.D. Tsadilas, and V. samaras. 2010. Trace elements availability in a sewage sludge-amended cotton grown Mediterranean soil. Chemosphere 80(11): 1308–1313).

Antoniou, G., N. Kathijotes, D.S. Spyridakis, and A.N. Angelakis. 2014. Historical development of technologies for water resources management and rainwater harvesting in the Hellenic civilizations. International Journal of Water Resources Development. 30(4): 680–693.

Apostolakis, C. G. and Douka C.E. 1970. Distribution of macro-and micronutrients in soil profiles developed on lithosequences and biosequences in Northern Greece. Soil Sci. Soc. Am. Proc. 34, 290–296.

Assimakopoulos, J.H., N.J. Yassoglou and C.P. Bovis. 1994. Effects of incubation at different water contents, air-drying and K-additions on potassium availability of a Vertisol sample. Geoderma 61: 223–236.

Bagnouls, F. et Gaussen, H. 1953. Saison seche et indice xerothermique. Doc. Carte Prod. Veget. III: 1–8.

Barbayiannis, N, V.P. Evangelou, and V.C. Keramids. 1996. Potassium-ammonium-calcium quantity/intensity studies in the binary and ternary modes in two soils of micaceous mineralogy of northern Greece. Soil Sci. 10: 716–724.

Bornovas, J., and Rontogianni-Tsampaou, Th. 1983. Geological map of Greece. Institute of Geological and Mineral Exploration. Athens, Greece.

CORINE, 1992. Soil erosion risk and important land resources in the southern regions of the European Community. EUR 13233, Luxembourg.

Dafis, S. 1966. Site and Forest Productivity Studies in Cooppice oak Forests and Chestnut Groves in Northeastern Chalkidiki. Lectурship Thesis. Laboratory of Silviculture and Mountain Hydronomy. Aristotelian University of Thessaloniki. Thessaloniki, Greece. 120pp (in Greek).

Dale, T. and V.CG. Carter. 1955. Topsoil and Civilization. University of Oklahoma Press. Pp. 174.

Danalatos, N.G, Kosmas, C.S., Moustakas, N. and Yassoglou N. 1995. Rock fragments II. Their impact on soil physical properties and biomass production under Mediterranean conditions. Soil Use and Management J. 11: 121–126.

Dionisiou, N., T. Matsi, and N. Misopolinos. 2006. Use of magnesia for B removal from irrigation water. Journal of Environmental Quality. 35(6): 2222–2228.

Drouza, S., F.A. Georgoulias, N.K. Moustakas. 2007. Investigation of soils developed on volcanic materials in Nisyros island, Greece. Catena 70: 340–349.

Economou, A., Michopoulos, P. and Nakos, G. 1993. The forest soils of the Western Macedonia Region: 2. Chemical properties. Proceedings of the 4th Greek soil conference on soil and environment. Greek So. Soc. Thessaloniki Greece. 587–595 (in Greek).

Emberger, L. 1955. Une classification biogeographique des climat. Rev.Trav. Fac. Sci. Montpellier, Bot. 7: 3–43.

Evangelou, E., P. Dalias, C. Giourga, and C. Tsadilas. 2014. Effect of land use history on soil carbon and nitrogen in a Mediterranean catchment. Commun. Soil Sci. Plant Anal. 45: 2331–2340.

F.A.O. 1976. A Framework for Land Evaluation. Soils Bulletin 32. FAO, Rome, 72pp.

FAO/ISRIC/ISSS 1998. World Reference Base for Soil Resources. World Soil Resources Report #84. FAO Rome 88pp.

Foxhall, L. 2006. Environments and Landscapes of Greek Culture. In K.H. Kinzl (Ed.) A Companion to the Classical Greek World. Blackwell Publishing pp. 245–28.

Foxhall, L. 2007. Olive Cultivation in Ancient Greece. Seeking the Athenian Economy. Oxford Univ. Press. pp. 313.

Gakis, S.F., M.Z. Orfanoudakis, A.G. Papaioannou, K.T. Mantzanas, V.P. Papanastasis, D.A. Alifragis, D.G. Seilopoulos, and S.N. Kostakis. 2014. Long term evolution of tree growth, understorey vegetation and soil properties in a silvopastoral system of northern Greece. Annals of Forest Research 57(2): 247–265.

Georgoulias, F. 2008. Genesis of volcanic soils in Greek conditions. Doctor's Thesis. Agricultural University of Athens (in Greek).

N. Yassoglou et al., *The Soils of Greece*, World Soils Book Series,
DOI 10.1007/978-3-319-53334-6

Georgoulias, F.A. and N.K. Moustakas. 2010. Exploration of soils developing on volcanic materials on the island of Milos, Greece. Catena 81: 43–54.

G.N.C.C.D. Greek National Committee for Combating Desertification, 2000. Greek national action plan for combating desertification, FEK 947/B' Ministerial Decision (in Greek).

Grove, A. T. and Rackham, Oliver, 2001. The nature of the Mediterranean Europe. An ecological History.Yale University Press, New Haven and London, 107–118.

Haidouti, C. and N. Yassoglou. 1982. The genesis of certain soil profiles on an archaeological site in southern Greece. SSSAJ. 46(5): 1047–1051.

Hanson, V.D. 1998. Warfare and Agriculture in Classical Greece. University of California Press. Pp. 359.

Hanson, V.D. 1995. The Other Greeks: The Family Farm and the Agrarian Roots of Western Civilation. University of California Press. Pp. 687.

Innocenti, F., Mannetti,P., Pecerillo, A., Poli, G. 1981. South Aegean Volcanic Arc: Geochemical Variations and Geotectonic Implications. Bulletin of volcanology. 44(3),377–391.

Isager, S. and J.E. Skydsgard 1995. Ancient Greek Agriculture. An Introduction. Routledge Publisher, London and New York. Pp. 245.

ISMC, Institute of Soil Mapping and Classification. 1989a. Soil Survey Study of the Region C of the West Part of the Zone Z of Acheloos River's Diversion Works. Pp. 86.

ISMC, Institute of Soil Mapping and Classification. 1989b. Soil Survey Study of the Region C of the Acheloos River's Diversion Works. Pp. 209.

ISMC, Institute of Soil Mapping and Classification. 1990a. Soil Survey Study of the Region C of the North Part of the Zone Z of Acheloos River's Diversion Works. Pp. 159.

ISMC, Institute of Soil Mapping and Classification. 1990b. Soil Survey Study of the Agia Region. Pp. 172.

ISMC, Institute of Soil Mapping and Classification. 1991a. Soil Survey Study of Perfecture of Karditsa. Pp. 522.

ISMC, Institute of Soil Mapping and Classification. 1991b. Soil Survey Study of Perfecture of Trikala. Pp. 137.

ISMC, Institute of Soil Mapping and Classification. 1995. Soil Survey Study of Perfecture of Pieria. Pp. 125.

ISMC, Institute of Soil Mapping and Classification. 1997. Soil Survey Study of Pinios Delta River of the prefecture of Larissa. Pp. 113.

ISST, Institute of Soil Science of Thessaloniki, 2006. Soil study of reducing inputs of chemical fertilizers and sound management of soil-water resources in the Municipality of Servia of the Prefecture of Kozani.

Jandi, R., M. Rodegniero, and M. Ollson, 2011. Soil carbon in sensitive European ecosystems: From science to land management. John Wiley & Sons, pp. 256).

Jenny, H. 1941. Factors of Soil Formation. A System of Quantitative Pedology. McGraw-Hill, Book Co., Inc. New York. 281 pp.

Jenny, H. 1980. The soil source; origin and behavior. Springer-Verlag. Berlin. 377 pp.

Kairis, O., C. Karavitis, A. Kounalaki, L. Salvari, and C. Kosmas. 2013. The effect of land management practices on soil erosion and land desertification in an olive grove. Soil Use Management 29(4): 597–606.

Kanellopoulos, C. A. Argyraki, and P. Mitropoulos. 2015. Geochemistry of serpentine agricultural soil and associated groundwater chemistry and vegetation in the area of Atalanti, Greece. Journal Geochemical Exploration. 158: 22–33.

Karamesouti, M. I. 2011. Definition of environmentally sensitive to desertification areas of Greece. Postgraduate Thesis. Agricultural University of Athens. {Ark}, 120 p.p. and Annexes (in Greek).

Karapiperis A. 1955. The climate of Greece. In the Grand Greek Encyclopedia, V: 10. Athens. (in Greek).

Karyotis, Th. K. 1994. Relation of soil nitrogen flow to the chemical and physical properties of the soil. Doctor's Thesis, Agricultural University of Athens. Athens pp. 199.

Karyotis, Th. Haroulis, A., Vavoulidou, E. and Papadopoulos, P. 2000. Soil properties and distribution of heavy metals and boron within three Greek Histosols. Suoseura-Finish Peatland Society, Suo(51 (3)), Helsinki pp. 95–104.

Kelepertzis, A., D. Alexakis, and K. Skordas. 2006. Arsenic, antimony and other toxic elements in the drinking water of Eastern Thessaly in Greece and its possible effects on human health. Environ. Geol. 50: 76–84.

Kelepertzis, E. 2014. Accumulation of heavy metals in agricultural soils of Mediterranean: Insights from Argolida basin Peloponnese, Greece. Geoderma (221-222): 82–90.

Kliengebiel, A.A. and Montgomery, P.H. 1961. Land capability classification. Agricultural Hand book 210. Washington DC: Department of Agriculture.

Kollias, V.J. Kalivas, D. P. and Yassoglou, N.J. 1999. Mapping the soil resources of a recent alluvial plain in Greece using fuzzy sets in a GIS environment. European Journal of Soil Science, June 50, 261–273.

Kosmas, C., Moustakas, N., Kallianou, C., Yassoglou, N. 1991. Cracking patterns, bypass flow and nitrate leaching in Greek irrigated soils Geoderma 46(4): 293–304.

Kosmas, C., Danalatos, N., Moustakas, N., Tsatiris, B., Kallianou, Ch., Yassoglou, N., 1993. The impacts of parent material and landscape position on drought and biomass production of wheat under semi-arid conditions. Soil Technology 6, 337–349.

Kosmas, C., Yassoglou, Moustakas, N., & Danalatos, N., 1995. Field site: Spata, Greece. In: Mediterranean Desertification and Land Use —MEDALUS II final report, Basic Field Programme. Commission of the European Communities, Contract Number EV5V-CT92-0128, pp. 129–163.

Kosmas, C., Moustakas, N., Danalatos, N. and Yassoglou, N., 1996. The Spata Field Site: I. The Impacts of Land Use and Management on Soil Properties and Erosion. II. The Effect of Reduced Moisture on Soil Properties and Wheat Production. In: J. Brandt, and J. Thornes [Eds], Mediterranean Desertification and Land Use. John Wiley & Sons, Chichester.

Kosmas, C., Danalatos, N., Cammeraat, L.H., Chabart, M., Diamantopoulos, J., Farand, R., Gutierrez, L., Jacob, A., Marques, H., Martinez-Fernandez, J., Mizara, A., Moustakas, N., Nicolau, J.M. Oliveros, C., Pinna, G., Puddu, R., Puigdefabregas, J., Roxo, M., Simao, A., Stamou, G., Tomasi, N., Usai, D., and Vacca, A. 1997. The effect of land use on runoff and soil erosion rates under Mediterranean conditions. Catena 29:45–59.

Kosmas, C., Briassoulis, H., Gerontidis, St., Detsis, V., Marathaianou, M., and Dalakou, B., 1998. Lesvos-soil vegetation interactions. In: Mediterranean Desertification and Land Use - MEDALUS III final report, Target areas. Commission of the European Communities, Contract Number EV4-CT95-0119, pp. 611–682.

Kosmas, C., M. Kirkby, and N. Geeson (Eds.). 1999. The Medalus project Mediterranean desertification and land use. Manual on key indicators of desertification and mapping environmentally sensitive areas to desertification. European Commission EUR 18882 pp. 94.

Kosmas, C., N. Danalatos, and St. Gerontidis. 2000. The effect of of land parameters on vegetation performance and degree of erosion under Mediterranean conditions. Catena 40: 3–17.

Kosmas, C., Gerontidis, S., Marathianou, M., 2000a. The effect of land use change on soils and vegetation over various lithological formations on Lesvos (Greece). Catena 40, 51–68.

Kosmas, C., N.G. Danalatos, and St. Gerondidis. 2000b. The effect of land parameters on vegetation performance and degree of erosion under Mediterranean conditions. Catena 40: 3–17.

Kosmas, C., and Antonoglou, N. 2002. Study of two water reservoirs and cental irrigation system in the municipalities of Farrton and

Olenias, Achaias: Soil survey report. Region of western Greece, Department of public works, direction of land management, 68 pp (in Greek).
Kosmas, C., Moustakas, N., Papageorgiou, G., Kosma, D., 2005. Protection of soils with the use of ecological engineering in river watershed level in the municipality of Filippon Kavalas. First annual report. Interreg IIIB Medocc Programme (IMONT) http://efunds.gov.mt/en, Municipality of Filippon, Kavala (in Greek).
Kosmas, C., Kairis, O., Kounalaki, A., Karamesouti, M. 2010. Study of soil characteristics for defining less favorable areas of Greece. Ministry of Rural Development and Foods, Agrogi AE. 20 pp. (in Greek).
Koukouzas, N.K., R. Zeng, V. Perdikakis, W. Xu, and E.K. Kakaras. 2006. Mineralogy and Geochemsitry of Greek and Chinese coal fly ashes. Fuel 85: 2301–2309.
Kyrkilis, D., S. Semasis, and C. Stliaras, 2014. The role of agriculture in economic growth in Greece. In A. Karasavvoglou and P. Polychronidoy (Eds.) Economic Crisis in Europe and Balkans, Contributions to Economics. DOI IO.1007/978-3-0319-00494-5_13).
Kyrkilis, D. And S. Semasis. 2015. Greek Agriculture's Failure. The other Face of the Failed Industrialization. From Accession to EU to the Debt Crisis. Procedia Economics and Finance 33: 64–77.
Lithourgidis, A.S., C.A. Damalas, and I.G. Eleftherochorinos. 2009. Conservation tillage: A promising perspective for sustainable agriculture in Greece. Journal of Sustainable Agriculture 33(1): 85–95.
Machura, M., and A. Mulawa. 1973. Rosenbrock function minimization. Algorithm 450. Comm. ACM 16:482–483.
Mariolopoulos, E. 1936. A cross section of meteorological data in Greece. Athens.
Mariolopoulos, E. 1938. The climate of Greece. Athens. (in Greek).,
Matsi, T. and V.Z. Keramidas. 1999. Fly ash application on two acid soils and its e€ect on soil salinity, pH, B, P and on ryegrass growth and composition. Environmental Pollution 104: 107–112.
Mavromatis, G. 1980. The Bioclimate of Greece, relations of climate and natural vegetation. Bioclimatic maps. Institute of Forest Research, Athens. 63pp and maps (in Greek).
Ministry of Agriculture, Agricultural University of Athens. 1983. Soil Survey Study of the Region of Agrolida Plain. Pp. 250 (in Greek).
Minstry of Agricultrure. 2008. The Greek agricultural sector in numbers. Basic characteristics. Athens. Pp. 291 (in Greek).
Mizota, C., Kusakabe, M. and Noto, M. 1988. Eollean contribution to soil development on Cretaceous limestones in Greece by oxygen isotope composition of quartz. Geoderma. J. (Jpn), 22: 41–46.
Misopolinos, N. 1985. A new concept for reclaiming sodic soils with high-salt water. Soil science 140: 69–74.
Misopolinos, N.D. 1987. Reclamation of a fine clayey montmorillonitic soil with high-salt water under laboratory condition. Zeitschrift fur Pflanzenernahrung und Bodenkund 150(6): 395–399.
Misopolinos, N.D., N.G. Silleos, and K.P. Prodromou. 1988. Catena 15 (2): 127–136.
Mizota, C., I. Kawasaki, I., and T. Wakatsuki, 1988. Clay mineralogy and chemistry of seven pedons formed in volcanic ash, Tanzania. Geoderma 43(2–3): 131–141
Mpokos, J.P., D.M. Yiakoulaki, Z.A. Papazafeiriou, S. Sgardelis, D. Alifragis, and K. Papanikolaou. 2014. Herbage Production and Species Richness in Sub-alpine Grasslands of Different Soil Parent Material in Northern Greece. Journal of Mountain Science 11 (6):1579–1592.
Moudrakis, D. 1985. Geology of Greece. University Studio Press. Athens. 207pp.
Moustakas, N., Kosmas, C., Danalatos, N., and Yassoglou, N. Rock fragments I. 1995. Their effect on runoff, erosion and soil properties under field conditions, Soil Use and Management J. 11:115–120.
Moustakas, N.K. and F. Georgoulias. 2005. Soils develoed on the volcanic materials in the island of Thera, Greece. Geoderma 129: 125–138.
Moustakas, N.K. 2012. A study of Vertisol genesis in North Eastern Greece. Catena 92: 208–215.
Nakos, G. 1979. Forest soils of Greece: Physical, chemical and biological properties. Forestry Ecology and Management V: 2 35–51.
Nakos, G. 1983a. Land resources survey of Greece. Journal of Environmental Mangement 17, 153–169.
Nakos, G. 1983b. Forms of phosphorus in some Greek forest soils. Commun. Soil Sci. Plant Anal. 14(1): 89–99.
Nakos, G. 1984. Relationships between bioclimatic zones and lithology with various characteristics of forest soils in Greece. Plant and Soil 79: 101–121.
Nakos, G. 1987. Phosphorus adsorption by forest soils. Commun. Soil Sci. Plant Anal. 18(3): 279–286.
Nevros, K.I. and Zvorykin, I.A. 1937. The variety of Solonetz Red soils in the Vicinity of the Village of Markopoulo, Attica. Soil Sci. 43, 177–239.
Noulas, C., D. Vlachostergios, and C. Tsadilas. 2013. Assessment of Phosphorus Supply Rate to Common Vetch By Using Ion Exchange Membranes. Commun. Soil Sci. Plant Anal. 44 (1–4): 275–286.
OPEKEPE. 2014. Development of a unified geo-information system of soil data and delineation of agricultural zones of the country. The project was coordinated by the Aristotle University of Thessaloniki, 1860 pp. (in Greek).
Panagiotopoulos, K.P., C.P. Papadopouloy, and A. Hatjiioannidou. 1994. Compaction and penetration resistance of an Alfisol and Entisol and their influence on root growth of maize seedlings. Soil and Tillage Research 31(4): 323–337.
Papadopoulos, A., C. Prochaska, F. Papadopoulos, N. Gantidis, and E. Metaxa. 2007. Determination and evaluation of cadmium, copper, nickel, and zinc in agricultural soils of western Macedonia, Greece. Environ. Manage. 40: 719–726.
Papamichos, N. 1979. Use of a system specially developed for quick examination, description, classification and mapping of mountain areas (Pieria Region). Ministery of Agriculture, Athens. 140pp {in Greek}.
Pelisek, J. 1964. Braune Wald Boden und ihre Degradation im Gebiete der der Tschechoslowakischen Republik. Int. Congr. Soil; Sci. Trans. 8th V: 393–399.
Polyzopoulos, N.A., V.Z. Keramidas and H. Kiosse. 1985. Phosphate sorption by some Alfisols of Greece as described by commonly used isotherms. SSSAJ. 49: 81–84.
Ponomayova, V.V. and Nikolayeva, T. A. 1964. Biochemical aspect of the genesis of Gray and Brown Forest Soils. Congr. Soil Sci. Trans. 8th V: 339–345.
Rubio, J. 2007. Soil Conservation: From Agricultural Productivity to Multifunctionality. ESSC Newsletter 2/2007. 3–10p.
Samaras, V., C.D. Tsadilas, and S. Stamatiadis. 2008. Repeated application of municipal sewage sludge in a cotton field: Effects on soil fertility, crop yield, and nitrate leaching. Agron. J. 100: 477–483.
Samaras, V., C.D. Tsadilas, S. Stamatiadis, and S. Nortcliff. 2010. Effects of biosolids from tomato processing on soil fertility and wheat growth in a Greek soil. Journal of Plant Nutr. And Soil Sci. 173(2): 252–259.
Sauvage, CH. 1963. Etages bioclimatiques. Atlas du Maroc. Notices explicatives. Comite National de Geographie du Maroc.Rabat.
Settas, N. 1975. Greece and its climate. The Athens Academy. 447pp. (in Greek).
Sgouras, I.D., C.D. Tsadilas, N. Barbayiannis, and N. Danalatos. 2007. Physicochemical and mineralogical properties of red Mediterranean

soils from Greece. Communications in Soil Science and Plant Analysis, Vol. 38:695–711.

Shahabi, S. 1982. The soils of Drama, with special reference to the soils of the Drama Basin (N. Greece). Institute for Chemistry and Agriculture, N. Canellopoulos. 178pp {in Greek}.

Shaheen, S.M., C.D. Tsadilas, and K. M. Eskridge. 2009. Effect of common ions on phosphorus sorption and lability in Greek Alfisols with different pH. Soil Sci. 174: 21–26.

Shaheen, S.M., Tsadilas, C.D., 2010. Sorption of cadmium and lead by acidic Alfisols as influenced by fly ash and sewage sludge application. Pedosphere 20: 436–445.

Shaheen, S.M. and C. Tsadilas. 2013a. Utilization of Biosolids in Production of Bioenergy Crops I: Impact of Application Rate on Canola Biomass, Soil Properties, and Nutrient Availability. Commun. Soil Sci. Plant Anal. 44(1–4): 243–258.

Shaheen, S.M., Tsadilas, C.D., 2013b. Phosphorus sorption and availability to canola grown on an Alfisol amended with various soil amendments. Commun. Soil Sci.Plan. 44: 89–103.

Shaheen, S.M., P. Hooda, and C.D. Tsadilas. 2014. Opportunities and challenges in the use of coal fly ash for soil improvements—A review. Journal Environmental Management. 145: 249–267.

Skodras, G., P. Grammedis, E. Kakaras, D. Karangelos, M. Ananostakis, and E. Hinis. 2007. Quality characteristics of Greek fly ashes and potential use. Fuel Processing Technology 88: 77–85.

Soil Survey Staff 1975. Soil Taxonomy. A Basic System of Soil Classification for Making and Interpreting Soil Surveys. US Government Print Office. Washington D.C, 20042. 754 pp.

Soil Survey Staff. 1992. Keys to Soil Taxonomy. SMSS Technical Monograph No 19 Fifth Edition. Pocahonas Press. Inc. Blacksbourg. Virginia. 541pp.

Soil Survey Division Staff. 1993. Soil Survey Mannual.USDA Handbook No 18. US Government Printing Office Washington DC20402. 437pp.

Sotiriou, G., Christodoulou, M. and Nakos G. 1993. The forest soils of the Western Macedonia Region: 1 Classification. Proceedings of the 4th Greek soil conference on soil and environment. Greek So. Soc. Thessaloniki Greece. 569–585 {in Greek}.

Suchkova, N., E. Darkas, and J. Ganoulis. 2010. Phytoremediation as a prospective method for rehabilitation of areas contaminated by long-term sewage sludge storage: A Ukrainian–Greek case study. Ecological Engineering 36, (4): 373–378.

Tsadila, E., L. Evangelou, C. Tsadilas, Ch. Giourga, and S. Stamatiadis. 2012. Land-Use Effect on Selected Soil Quality Parameters Commun. Soil Sci. Plant anal. 43: 595–604.

Tsadilas, C.D., N. Yassoglou, C. Kosmas, and Ch. Kallianou. 1994. Availability of soil boron fractions to olive trees and barley and their relationships to soil properties. Plant and Soil. 162:211–217.

Tsadilas, C.D., Th. Matsi, N. Barbayiannis, and D. Dimoyiannis. 1995. Influence of sewage sludge application on soil properties and on the distribution and availability of heavy metal fractions. Commun. Soil Sci. Plant anal. 26(15–16): 2603–2619.

Tsadilas, C.D., V. Samaras, and D. Dimoyiannis. 1996. Phosphate sorption by red Mediterranean soils from Greece. Commun. Soil Sci. Plant anal. 27(9–10): 2279–2293.

Tsadilas, C.D., D. Dimoyiannis, V. Samaras. 1997. Methods of assessing boron availability to kiwifruit plants growing on high boron. Commun.Soil Sci.Plant Anal. 28(11–12): 973–987.

Tsadilas, C.D. and V.P. Evangelou. 1998. Mixed soil/clinoptilolite quantity/intensity behavior in binary and ternary K-NH4-Ca. Soil Sci. 163: 454–462.

Tsadilas, C.D., D.G. Dimoyiannis, and V. Samaras. 1999a. Sewage sludge usage in cotton crop: I. influence on soil properties. Pedosphere 9(2): 147–152.

Tsadilas, C.D. and K.S. Chartzoulakis. 1999b. Boron deficiency in olive trees in Greece in relation to soil boron concentration. Acta Horticulturae. 474:341–344.

Tsadilas, C.D., Samaras, V., Kazai, P., Sgouras, J., 2002. Fly ash and sewage sludge application on an acid soil and their influence on some soil properties and wheat biomass production. In: Proceedings of the 2002, vol. III. International Soil Conservation Organization, Beijing, China, pp. 118–123. May 26–31, 2002.

Tsadilas, C.D. and P. Kazai. 2005. Evaluation of Methods for Determination of Soil Boron Available to Apple Trees. Commun. Soil Sci. Plant anal. 36(4–6): 695–700.

Tsadilas, C.D., Shaheen, S.M., Gizas, D., Samaras, V., Hu, Z., 2009. Influence of fly ash application on copper and zinc sorption by acidic soil amended with biosolids. Commun. Soil Sci. Plant Anal. 40, 168–179.

Tsadilas, C.D. and S.M. Shaheen. 2010. Distribution of total and ammonium bicarbonate-DTPA extractable soil vanadium from Greece and Egypt and their correlation to soil properties. Soil Sci. 175: 535–543.

Tsadilas, C.D. and S.M. Shaheen. 2013. Utilization of Biosolids in Production of Bioenergy Crops II: Impact of Application Rate on Bioavailability and Uptake of Trace Elements by Canola. Commun. Soil Sci. Plant Anal. 44(1–4): 259–274.

Tsadilas, C.D., M. Tziouvalekas, E. Evangelou, E., and A. Tsitouras. 2014a. Arsenic in groundwater, soils and crops of east side area of Ossa mountain, Greece. One Century of the Discovery of Arsenicosis in Latin America (1914–2014): As 2014—Proceedings of the 5th International Congress on Arsenic in the Environment pp 267–269.

Tsadilas, C., M. Tziouvalekas, E. Evangelou, and A. Tsitouras. 2014b. One century of the discovery of arsenicosis in Latin America (1914–2014): As 2014-Proceedings of the 5th International Congress on Arsenic in the Environment.

Tsatiris, B.D., Serelis, G., Kallianou, Ch.S. and Tsagalidis, A. 2008. Impact of serpentinization on soil properties and metal availagility. Catena. Advances in Ecology 39, pp 477–492.

Varelides, O., N. Yassoglou, and S. Liveris. 2004. Desertification: Greek Publications. Ministry of Rural Development and Foods. General Directorhsip of Development and Forest and Environment Protection. Pp, 774.

Wright, H. E. Jr. 1972. Vegetation history. In McDonald, W.A. and Rapp, G.R. Jr. (eds).The Minnesota Expedition. Reconstructing a Bronze Age Regional Environment. The University of Minnesota Press. Minneapolis, pp 188–199.

Yaalon, D.H. 1997, Comments on the source, transport and deposition scenario of Saharan dust to southern Europe. Journal of Arid Environments 36(1): 193–196.

Yassoglou, N. 1963. A classification of soil sites for poplars in Macedonia, Greece. Ministry of Agriculture, Forest Research.

Yassoglou, N. J., Henrard G. and Apostolakis C. 1963. First six months progress report of soil survey project of Pelopennese regional Development Service and Nuclear Research Center Democritus. Athens. Greece 35 pp. (in Greek).

Yassoglou, N.J., Catherine Nobeli and Vrachamis S. C. 1969. A Study of Some Biosequences and Lithosequences in the Zone of the Brown Forest Soils in Northren Greece. Morphological, Physical and Chemical Properties. S.S.S. Am. Proc. V: 33, no 291–296.

Yassoglou, N.J. 1971. Soil Science I. Description of the soil system. A text book for the students of the Agricultural University of Athens. 260 pp. (in Greek).

Yassoglou, N.J. and Catherine Nobeli. 1972. Soil Studies. In McDonald, W.A. and Rapp, G.R. Jr. (Eds).The Minnesota Expedition. Reconstructing a Bronze Age Regional Environment. The University of Minnesota Press. Minneapolis, pp 171–176.

Yassoglou. N.J. 1973. Soil Science II. The Dynamics and the Fertility of the Soil System. A text book for the students of the Agricultural University of Athens. 375pp. (in Greek).

Yassoglou, N.J., Nychas, A., Apostolakis, C and Kosmas K. 1983. Soil survey report of the Argos Plain. Ministry of Agriculture and Agricultural University of Athens. Greece 246 pp. (in Greek).

Yassoglloou, N.J. 1986. Production Potential Part II. Sensitivity of the soil systems in Southern Europe to degrading influxes. In Barth, H. and L' Hermite, P.L. eds. Scientific Basis for Soil Production in the European Community. Elsevier Applied Science London pp. 87–122.

Yassoglou, N., Kosmas, C., 1988. Potential erosion and desertification risks in Greece. Second Meeting of the Balkan Initiative Committee for the Balkan Scientific Conference on Environmental Protection in the Balkans, Sofia, 17–18 May 1988.

Yassoglou, N., Kosmas, C.S., Moustakasa, N., Tzianis, E., Danalatos, N.G. 1994. Cracking in recent alluvial soils as related to easily determined soil properties Geoderma 63(4):337–349.

Yassoglou, N.I. 1995. Lessons of Applied Soil Science. Agricultural University of Athens, pp. 234.

Yassoglou, N.J. 1996. Greece, Soil mapping and soil data bases In Le Bas and Jamagne, M. (eds) Soil databases to support sustainable development. Joint Research Centre, European Commission and INRA. EUR 16371 EN. 57–60p.

Yassoglou, N., C. Kosmas, and N. Moustakas. 1997. The red soils, their origin, properties, use and management in Greece. Catena 28: 261–278.

Yassoglou, N.J. 2004. Soil Associations Map of Greece. Greek National Committee for Combating Desertification and Agricultural University of Athens.

Yassoglou, N. 2015. A Critique of Strategies and Policies for Soil Consevation throu the Centuries. InFullen, M.A., Famodimou, J., Karyotis, Th., Noulas, Chr., Pagopoulos, A., Rubio, J.L. and Gabriels, D. (eds.) 2007. Inovative Styrategies and Land Policies for Soil ConsAervation. Advances in Geoecology 44. CATENA VERLAG GMBH 354547 Reiskirchen, Germany. pp. 1–10.

Zalidis, G., Stamatiadis, S., Takavakoglou, V., Eskridge, K. and Misopolinos N. 2002. Impacts of agricultural practices on soil and water quality in the Mediterranean region and proposed assessment methodology. Agriculture Ecosystems and Environment 88(2): 137–146.

Zvorikin I.,A. and Vrisimidji S. A. 1947. Soil Map of Thessaly. Institute of Chemistry and Agriculture. "N. Kanelopoulos" Soils Institute of the Ministry of Agriculture Athens, Greece.

Appreciations

The authors express their appreciation to all the authors, whose scientific work provided data used in this book. especially to Dr. George Nakos for his significant research contribution in the field of Forest Soils. Thanks are extended to A.Tsitouras, E. Evangelou, K. Ntokouzis T. Nicoli, amd M. Karamesouti who helped in the preparation of this book.

N. Yassoglou et al., *The Soils of Greece*, World Soils Book Series,
DOI 10.1007/978-3-319-53334-6

Index

N. Yassoglou et al., *The Soils of Greece*, World Soils Book Series,
DOI 10.1007/978-3-319-53334-6

MIX
Papier aus verantwortungsvollen Quellen
Paper from responsible sources
FSC® C105338

If you have any concerns about our products,
you can contact us on
ProductSafety@springernature.com

In case Publisher is established outside the EU,
the EU authorized representative is:
Springer Nature Customer Service Center GmbH
Europaplatz 3, 69115 Heidelberg, Germany

Printed by Libri Plureos GmbH
in Hamburg, Germany